$L7.$

D1135484

EXPLORATION OF
THE OUTER SOLAR SYSTEM

Edited by
Eugene W. Greenstadt
TRW Inc.
Redondo Beach, California

Murray Dryer
National Oceanic and Atmospheric Administration
Boulder, Colorado

Devrie S. Intriligator
University of Southern California
Los Angeles, California

Volume 50
PROGRESS IN
ASTRONAUTICS AND AERONAUTICS

Martin Summerfield, Series Editor-in-Chief
Princeton University, Princeton, New Jersey

Technical papers selected from AIAA/AGU Space Science
Conference: Exploration of the Outer Solar System, July 1973,
subsequently revised for this volume.

Published by the American Institute of Aeronautics and
Astronautics.

American Institute of Aeronautics and Astronautics
New York, New York

Library of Congress Cataloging in Publication Data
Main entry under title:

Exploration of the outer solar system.

   (Progress in astronautics and aeronautics; v. 50)
   Includes bibliographies and index.
   1. Outer space——Exploration——Congresses. 2. Solar wind——
Congresses. 3. Planets——Congresses. 4. Comets——Congresses.
I. Greenstadt, Eugene W. II. Dryer, Murray. III. Intriligator, Devrie S.
IV. AIAA/AGU Space Science Conference: Exploration of the Outer
Solar System, Denver, 1973. V. Series.
TL507.P75 vol. 50 [QB501] 629.1'08s [523] 76-54804
ISBN 0-915928-14-0

Table of Contents
Volume 50

Progress in
Astronautics and Aeronautics

Martin Summerfield,
Series Editor
PRINCETON UNIVERSITY

VOLUMES

EDITORS

1. Solid Propellant Rocket
Research. 1960

Martin Summerfield
PRINCETON UNIVERSITY

2. Liquid Rockets and
Propellants. 1960

Loren E. Bollinger
THE OHIO STATE UNIVERSITY

Martin Goldsmith
THE RAND CORPORATION

Alexis W. Lemmon Jr.
BATTELLE MEMORIAL INSTITUTE

3. Energy Conversion for
Space Power. 1961

Nathan W. Snyder
INSTITUTE FOR DEFENSE ANALYSES

4. Space Power Systems. 1961

Nathan W. Snyder
INSTITUTE FOR DEFENSE ANALYSES

5. Electrostatic Propulsion. 1961

David B. Langmuir
SPACE TECHNOLOGY LABORATORIES, INC.

Ernst Stuhlinger
NASA GEORGE C. MARSHALL SPACE
FLIGHT CENTER

J. M. Sellen Jr.
SPACE TECHNOLOGY LABORATORIES

6. Detonation and Two-Phase
Flow. 1962

S. S. Penner
CALIFORNIA INSTITUTE OF TECHNOLOGY

F. A. Williams
HARVARD UNIVERSITY

7. Hypersonic Flow Research.
1962

Frederick R. Riddell
AVCO CORPORATION

8. Guidance and Control. 1962

Robert E. Roberson
CONSULTANT

James S. Farrior
LOCKHEED MISSILES AND SPACE
COMPANY

9. Electric Propulsion
Development. 1963

Ernst Stuhlinger
NASA GEORGE C. MARSHALL SPACE
FLIGHT CENTER

(Other volumes are planned.)

PREFACE

This volume reviews the status of some principal areas of current scientific interest in man's early exploration of the outer solar system.

As a matter of history, it is of interest to note the relative scarcity of this type of book in the AIAA Progress in Astronautics and Aeronautics series. This is only the fifth volume in the Series to deal chiefly or wholly with the physical universe that man investigates* rather than with the technology by which he conducts his investigations. We observe that all the "environmental" books are relatively recent, and we think this significant. The early volumes, beginning in 1960, followed a clearly defined national commitment to move out into space as rapidly as possible and dealt fittingly with the technological means of getting there. Now, with the first space probe and spacecraft accomplishments behind us, public interest is flickering, so that reviews of what we have learned scientifically and discussions of what we hope to learn in the future are appropriate at this time to provide the fuel for rekindling the fires of curiosity that drive engineering technology.

Nowhere is the tinder of inquiry more likely to be reignited than among the outer planets. Indeed, the argument is often heard today in the space science community that a thorough knowledge of the field and particle processes at work in the plasma of the mighty Jovian magnetosphere will measurably improve our understanding of similar processes in the much humbler, but still inadequately comprehended, magnetosphere of the Earth Such an understanding also could be applied to various plasma physical phenomena in the inner solar system, at the sun, and at certain types of other more distant stars. To those acquainted with history, this argument is no innovation. Rather it represents a very conservative view, hallowed by long and profitable experience. On a January night in 1610, Galileo pointed his telescope at Jupiter and observed for the first time four small bodies circling the planet. His observation that four Jovian satellites behaved according to the model of the sun and planets suggested by Copernicus was the first powerful argument for validity of the heliocentric Copernican system and may be credited with responsibility for firmly establishing the modern scientific, social, and theological revolution of which we are the inheritors. Thus, it was a telescopic exploration of the outer solar system that produced our present concept of the inner solar system and determined a good many modern notions about the inner man as well.

Although we still believe the outer solar system is the place to look for solutions to some local problems, our methods have changed. We no longer depend on reflected light and Earth-bound telescopes to define our planetary horizons. We can now send instrumented probes into space equip-

---

*We refer to Volumes 22, 1969; 27, 1972; 28, 1972; and 30, 1972. The titles are on pp. v-x of this volume.

ped to measure almost every material and ethereal local variable with arbitrary precision, limited only by our imaginations and our budgets. When we use the term "exploration" here, we therefore mean the dispatch of spacecraft on lengthy excursions covering tens of astronomical units (a.u.) and lasting several years. As for the term "outer," at the instant of geological time when the contents of this book are being prepared, edited, and printed, the outer solar system is thought of as the volume of space contained in the spherical shell centered at the sun and extending from outside the orbit of Mars to the aphelion distance of the furthermost elliptical comet trajectory, somewhere beyond 10^{14} km.

Our defined outer solar system is not a negligible volume of space by terrestrial standards, since it measures at least 10^{18} a.u.^{3} and quite possibly a few orders of magnitude more. Symmetry arguments mercifully excuse us from any compulsion to explore the whole shell, whereas native thrift prevents the immediate dispatch of instrumented probes to every known object of potential interest. Attention focuses therefore on a few selected items of particular importance. At this initial stage of exploration, the items that command attention are the major planets, the comets, and the medium through which these bodies travel.

An exploration strategy has been outlined by NASA, for which the major milestones over the next 15 years may be summarized as follows:

Target	Launch date	Target date	Objective
Jupiter	1977	1979 flyby	Improved survey of field, particle, and compositional properties of Jovian magnetosphere and atmosphere, plus imaging.
Saturn (MJS)	1977	1981 flyby	Same as above
Uranus	1977	1984 flyby	Optional target of preceding MJS 77 mission to make first-order survey of near-Uranus environment.
Jupiter	1981/82	1985 entry	Determine composition and physical properties of Jovian atmosphere.
Jupiter	1981/82	1986 orbiter	Map magnetosphere and inspect Jovian satellites.
Giacobini-Zinner	1984	1985 flyby	Detection and characterization of cometary nucleus and coma, and interaction of coma with solar wind.
Halley	1984	1986 flyby	Same as above.
Uranus	1984	1991 entry	Determine first-order composition and physical properties of atmosphere.
Saturn	1985	1987 entry	Same as above.

In addition to these targets and objectives, the solar wind will continue to be a subject of attention in continuing data acquisition by Pioneers F and G and in the transit phases of each mission listed. Also, tentative plans for a probe to orbit Saturn and land on Titan suggest a future as full of excitement and new discovery as the past has been. It may be anticipated that the kind of ambition, patience, and thinking represented by such a long-term strategy, if coupled with the necessary commitment to carry it out, already signals a revolutionary maturity in human affairs.

The original source of the collection of papers in this volume was the AIAA/AGU Space Science Conference: Exploration of the Outer Solar System, held in Denver, Colo., in July 1973. Papers were selected from those presented at the meeting, brought up to date as dictated by later events, and supplemented by appropriate additional contributions to make a compact picture of our chosen topics.

The first key element in this aspect of space exploration is the extended heliosphere itself, for the sun projects its material presence far beyond the inner planets by virtue of the constantly streaming, hypersonic solar wind. Study of the solar system, or of the sun as a star, is therefore incomplete without a comprehensive picture of the heliosphere all the way to its boundary with the interstellar medium. The wind is believed to have its boundary—i.e., its transition from solar-generated to interstellar gas—somewhere in the "outer" region defined above. This is the subject of the first group of papers.

Whereas the sun and heliosphere constitute the hot material of the solar system, the planets, their satellites, and various minor bodies constitute the cold, or condensed, material of the solar system. In this category, Jupiter and Saturn are the most important of the bodies in regular orbits. These two major planets account for 92% of the condensed mass in the solar system. Moreover, their nonmaterial extensions into the uncondensed solar wind are by no means negligible. The magnetosphere of Jupiter, for example, is several times the diameter of the sun and is alone the largest entity in the solar system except for the heliosphere itself. The second group of papers concentrates on the giant planets and their immediate environments.

Although the prospect of sending probes directly to bodies at, let alone beyond, the visible perimeter of the solar system is dim in the immediate future, the distant solar system generously sends representatives inward to us so that, if we wish to know something about the matter of which the far region is composed, we need only intercept one of these messengers with one of our own. These samples from the remote reaches of solar gravity are the comets, whose exploration constitutes one of the most rewarding of prospective new endeavors. The third group of papers deals with this topic.

There remains only the pleasure of expressing our gratitude to the many people who were involved in large or small ways in organizing the Conference and in preparing this volume. The original Conference was arranged by Rolf Faye-Petersen, then Chairman of the AIAA Technical

Committee on Space and Atmospheric Physics. He was succeeded in that position by Kenneth Moe, whose continuing encouragement was invaluable. We are indebted to Bruce Whitehead, Ray L. Newburn, Stephen F. Sousk, and James B. Weddell for important consultations in realizing this volume and assembling the contributions, and to a group of anonymous reviewers who gave us their constructive and unselfish assistance. The advice and hard labor of Ruth F. Bryans, AIAA Director, Scientific Publications, and Martin Summerfield, Series Editor-in-Chief, were irreplaceable in achieving publication, and the aid of Jeanne Graham and Marti Neale was indispensable in handling the editorial tasks.

Eugene W. Greenstadt
TRW Inc.

Murray Dryer
National Oceanic and Atmospheric Administration

Devrie S. Intriligator
University of Southern California

October 1976

CHAPTER I—SOLAR WIND

THE INTERPLANETARY MAGNETIC FIELD:
ITS EFFECTS ON THE SOLAR WIND FLOW

Paul ͷ. Coleman Jr.*

University of California, Los Angeles, Calif.

Abstract

Planetary spacecraft, deep space probes, lunar satellites,
and high-altitude Earth satellites have provided measurements of
the solar wind and the interplanetary magnetic field over a significant fraction of the time since the flight of Mariner 2 to
Venus in 1962. We now have a rather good description of the
typical state of this tenuous, magnetized plasma in the near-
Earth region of interplanetary space, and current work includes
studies of variations with time and location in space. Some of
the recent developments of such studies of the interplanetary
magnetic field and their interpretation in terms of solar, or
stellar, processes and the behavior of astrophysical plasmas
are discussed here. More specifically, models of the electric
current in the solar wind are developed, and the effects of the
resulting electromagnetic forces upon the solar wind phenomena
observed in a limited region of interplanetary space. The current consists of two components. The density of one depends
only upon the sun's dipole moment, and that of the other depends
upon the solar wind velocity and the sun's angular velocity as
well as its dipole moment. The latter component flows in heliographic meridional planes. The electromagnetic forces of this
component tend to accelerate the plasma in the leading half of

---

Presented as Paper 73-557 at the AIAA/AGU Space Science
Conference: Exploration of the Outer Solar System, Denver,
Colo., July 10-12, 1973. This work was supported in part by
NASA under Research Grants NGR 05-007-065 and NGL 05-007-004.
A portion of the computer costs was covered by The Regents of
the University of California.

*Professor of Planetary Physics, Department of Geophysics
and Space Physics, and Institute of Geophysics and Planetary
Physics.

3

a magnetic sector and to decelerate it in the following half.
If the angle between the sun's spin axis and the dipole axis is
different from 0° or 90°, the electromagnetic force near the
equatorial plane has a southward component in the positive sector and a northward component in the negative sector.

Introduction

The continuous emission by the sun of the ionized plasma
that forms the solar wind, because of the high electrical conductivity and high flow speed of the plasma, extends the sun's
magnetic field into interplanetary space. Because of the sun's
rotation, this extension of the sun's magnetic field involves
continuous losses of electromagnetic energy and momentum by the
sun. Although the loss rates associated with these electromagnetic components are small compared to those associated with
other processes, they are nevertheless significant because they
account for about half of the losses of rotational energy and
angular momentum sustained by the sun. The resulting electromagnetic braking of the sun's rotation has been treated in a
number of papers (e.g. Refs. 1-5).

Most models of the solar wind flow, after the example set
by Parker[6] in his pioneering work on the subject, do not include the electromagnetic forces exerted on the solar wind plasma because of the sun's rotation and its magnetic field. Here
we shall discuss some of the properties of these electromagnetic forces for some simple models of the sun's magnetic field
and the results of some calculations of their effects on the
solar wind flow.

These results indicate that the effects should be measurable with existing spacecraft instruments. To the extent that
such models satisfy the constraints imposed by the empirical
observations in the limited region of interplanetary space accessible to spacecraft, they will provide a base for extrapolation to other regions, better understanding of the solar wind
and its effects on the sun, and, finally, further insight into
stellar processes in general.

A Model of the Electric Current in the Solar Wind

We assume that the solar wind flows according to Parker's[6,7]
model, so that the meridional component of the velocity is
strictly radial. We employ a spherical polar coordinate system
(r, θ, ϕ) with polar axis parallel to the sun's spin axis. In
this system, the velocity $\vec{v}$ in Parker's model has only r and ϕ
components, with v_r approaching a constant value and v_ϕ appro-
ching zero with increasing distance from the sun. Thus, in

this model the electromagnetic forces and rotational effects on the solar wind flow are neglected. Furthermore, the lines of magnetic force are assumed to follow the streamlines of the solar wind velocity as described in the frame of reference rotating with the sun.

The spherical polar coordinates in the rotating frame (r', θ', ϕ') are related to those in the inertial frame (r, θ, ϕ) as follows:

$$r = r', \quad \theta = \theta', \quad \phi = \phi' + \Omega t$$

where Ω is the angular velocity of the rotating system, and t is the time. Since the magnetic field in the solar wind plasma follows the streamlines as they appear in the rotating frame, the field is

$$B_{r'} = B_{or'} \left[r_0', \theta', \phi' + (\Omega/v_r)(r - r_0) \right] (r_0'/r')^2$$

$$B_{\theta'} = 0$$

$$B_{\phi'} = -B_{r'}(\Omega r/v_r) \sin \theta'$$

for $r' > r_0'$. Here $r' = r_0'$ defines the "source" surface for the solar wind, $\vec{B}_0$ is the magnetic field at this surface, and it has been assumed that $|v_\phi'| << |\Omega r'|$ for $r' \geq r_0$.

The time dependence in the inertial frame of this spiral magnetic field of Parker may be obtained explicitly through the coordinate transformation just given. Thus,

$$B_r(r, \theta, \phi, t) = B_{or}(r_0, \theta, \alpha)(r_0/r)^2$$

$$B_\theta = 0$$

$$B_\phi = -B_r(\Omega r/v_r) \sin \theta$$

where $\alpha = \phi - \Omega t + (\Omega/v_r)(r - r_0)$.

To describe B_{or}, we require a model for the sun's magnetic field. For this model, we shall employ a dipole at an angle λ to the sun's axis of rotation, which is the polar axis of our coordinate system. At time $t = 0$, we assume that the dipole axis lies in the meridian plane defined by $\phi = 0$. Then, at $t = 0$, this dipole field has components

$$B_r = 2(a/r^3)(\cos \lambda \cos \theta + \sin \lambda \sin \theta \cos \phi)$$

$$B_\phi = (a/r^3)(\cos \lambda \sin \theta - \sin \lambda \cos \theta \cos \phi)$$

$$B_\phi = (a/r^3) \sin \lambda \sin \theta$$

Thus, for $B_{or}(r_0, \theta, \phi, t)$ we have

$$B_{or} = 2(a/r^3)[\cos \lambda \cos \theta + \sin \lambda \sin \theta \cos (\phi - \Omega t)]$$

and for the spiral field in the solar wind we have the components given in Table 1.[8]

To describe the electric current density $\vec{J}$ in the solar wind, we use, from Maxwell's equations in vacuuo,

$$\vec{\nabla} \times \vec{B} = (4\pi/c) \, \vec{J} + (1/c) \, (\partial E/\partial t)$$

and neglect the displacement current. From the expression for the spiral field $\vec{B}$, the components of $\vec{J}$ are those given in Table 1. It is shown easily that the displacement current is negligible.

From these expressions, it is apparent that there are two modes of current generation. One is a consequence of the forced motion of the solar plasma through the solar magnetic field, and the other is a consequence of the sun's rotation and unipolar induction. For the former, the strength of the current is independent of the sun's angular velocity. It is shown easily that these two parts of the current separately satisfy $\vec{\nabla} \cdot \vec{J} = 0$.

Here we see that, if $\lambda = 0$, the current is entirely radial and the density is proportional to $(r_0/r)^2$. Furthermore, for a dipole transverse to the spin axis, i.e., for $\lambda = \pi/2$ or for any $\lambda \neq 0$, the current produced by the sun's rotation is confined strictly to meridional planes ($\phi = $ const). The current pattern in a meridional plane is shown schematically for $\lambda = 0$ in Fig. 1, that for $\lambda = \pi/2$ in Fig. 2, and that for $\lambda = \pi/4$ in Fig. 3.[8] It should be emphasized that the streamline of J_i, the component of $\vec{J}$ produced by the sun's rotation, in a meridian plane is not a projection of the streamline onto that plane.

With $\vec{B}$ and $\vec{J}$ given, the electromagnetic body forces, $\vec{F} = (1/c) \, \vec{J} \times \vec{B}$, in the solar wind flow may be computed. The resulting expressions for the components of the force are listed in Table 1. These forces are neglected in Parker's model for the flow. Our purpose here is to describe qualitatively some of the likely effects of these forces, and our concern is primarily with $\vec{J}_i$, the rotationally induced component of the current.

Let us first consider the case $\lambda = \pi/2$, so that the sun's magnetic dipole lies in the sun's equatorial plane perpendicu-

Table 1 Magnetic field, electric current, and electromagnetic force for dipolar field

$B_r = B_0(r_0/r)^2$ (cos λ cos θ + sin λ sin θ cos α)

$B_θ = 0$

$B_φ = -B_r(\Omega r/v_r)$ sin θ $= -B_0(r_0/r)^2(\Omega r/v_r)$

where α = φ - Ωt + Ω(r - r_0)/v_r

$J_r = -(c/4\pi)B_0(r_0/r)^2(1/r)(\Omega r/v_r)[3 \sin λ \sin θ \cos θ \cos α + \cos λ (2 \cos^2 θ - \sin^2 θ)]$

$J = -(c/4\pi)B_0(r_0/r)^2(1/r) \sin λ \sin α [1 + (\Omega r/v_r)^2 \sin^2 θ]$

$J = -(c/4\pi)B_0(r_0/r)^2(1/r)(\sin λ \cos θ \cos α - \cos λ \sin θ)$

$F_r = (1/4\pi)B_0^2(r_0/r)^4(1/r) [\sin^2 λ \sin α \cos α (\Omega r/v_r) \sin^2 θ$
$+ \sin λ \cos λ \sin α (\Omega r/v_r) \sin θ \cos θ] [1 + (\Omega r/v_r)^2 \sin^2 θ]$

$F_θ = (1/4\pi)B_0^2(r_0/r)^4(1/r) \{\sin^2 λ \cos^2 α \cos θ \sin θ[-1 - 3(\Omega r/v_r)^2 \sin^2 θ]$
$+ \cos^2 λ \sin θ \cos θ [1 - 2(\Omega r/v_r)^2 + 3(\Omega r/v_r)^2 \sin^2 θ]$
$+ \sin λ \cos λ \cos α [-1 + 2 \sin^2 θ - 5(\Omega r/v_r)^2 \sin^2 θ + 6(\Omega r/v_4)^2 \sin^4 θ]\}$

$F = (1/4\pi)B_2^2(r_{0_2}/r)^4(1/r)[\sin^2 λ \sin α \cos α \sin θ + \sin λ \cos λ \cos θ]$
$[1 + (\Omega r/v_r)^2 \sin^2 θ]$

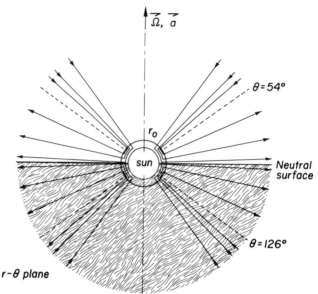

Fig. 1 Sketch of some streamlines of the component of the solar wind electric current generated by the sun's rotation for the case in which the sun's magnetic field is dipolar and the axis of the dipole is parallel to the axis of rotation of the sun (the case $\lambda = 0$). The shaded and unshaded areas are the meridional cross sections of regions in which the magnetic field is, respectively, toward or away from the sun.

lar to the spin axis, and the pattern of rotationally induced current is something like that sketched in Fig. 2. In the meridional cross section, the electromagnetic force $(1/c)(\vec{J_i} \times \vec{B})$ is directed away from the center of each field sector toward its boundary. Thus, if this force is not balanced by a pressure gradient, the plasma will flow away from the center toward the boundary as it moves away from the sun, and the density and field strength, to the extent that the magnetic field is frozen into the plasma, will tend to decrease near the center of a sector.

If the field sector is constrained by adjacent field sectors from expanding to accommodate this flow, then the density and field strength will increase toward the boundaries as they decrease at the center. Thus, the gradient in the magnetic field strength at the boundary of a field sector will increase relative to that indicated by the expressions for $\vec{B}$ in Table 1.

To expand somewhat on this point, the expressions for $\vec{B}$ in Table 1 indicate that an observer at a particular distance

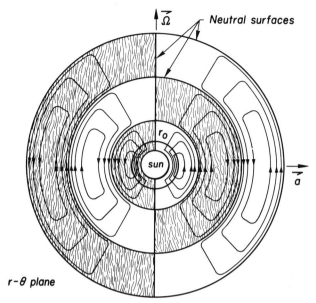

Fig. 2   Same as Fig. 1 for the case in which the dipole axis is perpendicular to the sun's spin axis (the case $\lambda = \pi/2$).

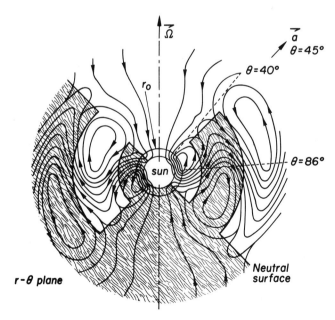

Fig. 3   Same as Fig. 1 for the case in which the dipole axis is inclined 45° to the sun's spin axis (the case $\lambda = \pi/4$).

from the sun (at a position moving, but not rotating, with the sun) will measure vector field components that alternate sinusoidally. The expression for F_r, the radial component of the force $(1/c)$ $\vec{J}_i$ x $\vec{B}$, indicates that $F_r > 0$ (away from the sun) from the time the polarity reverses until the absolute field strength reaches a maximum, and $F_r < 0$ from that time until the time of the next reversal. Thus, F_r is directed away from the center of the magnetic field sector toward the boundaries at which the field reverses polarity. This effect is simply a consequence of the tendency for the confined tube of magnetic flux to expand. These forces also will tend to expand the flux tube in the θ direction.

For the case $\lambda = 0$, the electromagnetic body force has only a θ component given by

$$F_\theta = (1/4\pi)(2a/r_0^3)^2(r_0/r)^4(1/r) \cos^2 \lambda \sin \theta \cos \theta$$

$$[1 - 2(\Omega r/v_r)^2 + 3(\Omega r/v_r)^2 \sin^2 \theta]$$

From this expression, we see that for $\Omega = 0$ and $\lambda = 0$ the body force F_θ is produced by $\vec{J}$ and is everywhere toward the equatorial plane. However, for $\Omega \neq 0$ there is a contribution to F_θ from the rotationally induced current, and this is away from the equator at colatitudes in the ranges 0°-54° and 126°-180° and toward the equatorial plane in the range 54°-126°.

Thus, for $(\Omega r/v_r) > 0$, the colatitude range over which the net electromagnetic force is toward the equator decreases with increasing distance from the sun and asymptotically approaches the range 54°-126°. For example, at 1.0 a.u. with $v_r = 450$ km/sec, $(\Omega r/v_r) \simeq 1$, and this range is 35°-145°. At 2.0 a.u. it is 50°-130°. If Ω were doubled, these last two ranges would apply at 0.5 and 1.0 a.u., respectively.

For the case $\lambda = 0$, then, the electromagnetic forces of the unipolar induction current tend to reduce the plasma density and magnetic field strength at midlatitudes and to increase both quantities at high and low latitudes relative to their values for radial flow. The radial dependences are not affected, to first order, because the electromagnetic force has no radial component when the dipolar and rotational axes are coincidental. For values of λ between 0° and 40°, the situation is more complicated. However, an important feature of the electromagnetic force in this case is the presence of a component southward across the equatorial plane where the field is directed away from the sun and a northward component where it is directed toward the sun.

Effects on the Flow: Some Quantitative Results

The radial solar wind flow, or more specifically flow
with $v_\theta = 0$, is not consistent with the presence of a solar
magnetic field. In order to determine how such forces will
affect the flow quantitatively, it is necessary to work the
flow problem with these forces included.

A few cases with azimuthal symmetry have been treated
numerically.[9-12] The steady-state equations of motion
were solved under the assumptions of perfect electrical conductivity, a polytropic relation between the pressure and density, and negligible viscosity. The total force on the plasma
is then the negative gradient of the pressure and the gravitational, rotational, and electromagnetic forces. The boundary
conditions specified at the base of the corona are the temperature, density, and magnetic field strength. For comparison
with the functions listed in Table 1 for a dipolar magnetic
field, expressions for the solar wind magnetic field, electric
currents, and electromagnetic body force are shown in Table 2
for the case in which the field strength at r_0 is constant over
each of the northern and southern hemispheres, directed away
from the sun in the north and toward the sun in the south. As
before, the functions listed in Table 2 were obtained under
the assumption that $v_\theta = 0$, i.e., that the solar wind flows
according to Parker's model.

The three critical surfaces associated with the flow equations supply the other three of the required six boundary
conditions. It also is assumed that the magnetic field and
the flow velocity are parallel in the reference system rotat-
ing with the sun. Thus, $\vec{B} = \kappa\rho\vec{v'}$, where ρ is the density and
$\vec{v'} = \vec{v} - \Omega r \sin\theta$. The physical significance of the parameter
κ is apparent from the requirement that $\kappa^2\rho/4\pi = 1$ at the Alfven
point in the flow or $\kappa^2 = 4\pi/\rho_A$, where ρ_A is the density
of the Alfven point.

The equations then are expanded about the radial, nonrotating solution of Parker, and an analytic expression is obtained for the resulting first-order equations, in terms of a
one-dimensional radial differential equation that is integrated easily by machine.

The expansion parameter $\epsilon = (\omega_s r_{Ap}/V_{Ap})^2$ is obtained by
including the Parker magnetic field in the momentum equation,
with the assumption that the flow properties remain unchanged.

Table 2   Magnetic field, electric current, and electromagnetic force for field of constant strength at $r = r_0$

$$B_r = (B_0/\beta)(r_0/r)^2$$

$$B_\theta = 0$$

$$B_\phi = -B_r(\Omega r/v_r) \sin \theta = -(B_0/\beta)(r_0/r)^2(\Omega r/v_r) \sin \theta$$

where $\beta = [1 + (\Omega r_0/v_r)^2 \sin^2 \theta]^{1/2}$.

$$J_r = (c/4\pi)\{(-2B_0/\beta)(r_0/r)^2(1/r)(\Omega r/v_r) \cos \theta$$
$$+ (B_0/\beta^3)(r_0/r)^2(1/r_0)(\Omega r_0/v_r)^3 \sin \theta \cos \theta\}$$

$$J_\theta = 0$$

$$J_\phi = (c/4\pi)(B_0/\beta^3)(r_0/r)^2(1/r)(\Omega r_0/v_r)^2 \sin \theta \cos \theta$$

$$F_r = 0$$

$$F_\theta = (1/4\pi) \{-2(B_0{}^2/\beta^2)(r_0/r)^4(1/r)(\Omega r/v_r)^2 \sin \theta \cos \theta$$
$$+ (B_0{}^2/\beta^4)(r_0/r)^4 (1/r_0)(\Omega r/v_r)(\Omega r_0/v_r)^3 \sin^2 \theta \cos \theta$$
$$+ (B_0/\beta^4)(r_0/r)^4 (1/r) (\Omega r_0/v_r)^2 \sin \theta \cos \theta$$

$$F_\phi = 0$$

The neglected magnetic energy per unit mass at infinity then is given by $(V_{Ap}/V_p) \varepsilon \sin^2 \theta$, and the effects of the neglected rotational magnetic field are expanded in terms of this parameter. Note that ε becomes zero if either the rotation rate or the field is zero.

In Table 3, the values of several pertinent variables are listed for various latitudes at 1.0 a.u. The two cases, Tables 3a and 3b, are different sets of boundary conditions at the corona. In Parker's model, v_θ and B_θ are zero. Thus, the different behaviors of these two variables in this self-consistent solution are readily apparent. The effects on v_θ are shown graphically in Figs. 4a and 4b, and the deviation of the flow streamlines from surfaces of constant θ is shown in Fig. 5.

Latitudinal Variations in the Coronal Boundary Conditions

Next, the equations were generalized further to permit first-order latitudinal variations in the specified coronal

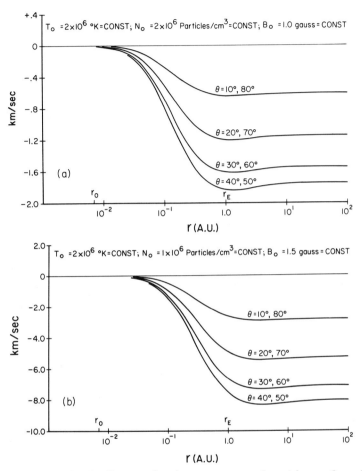

Fig. 4   Latitudinal flow velocity v_θ as a function of radial distance r for various choices of colatitude θ.   a) ε = 0.03, b) ε = 0.12.

boundary conditions in terms of Legendre polynomials at the boundary surface.   The lowest-order latitudinal variation with symmetry across the equatorial plane was treated.   Thus, the boundary conditions took the forms

$$T_0 = T_{op} \left(1 + a_1 \varepsilon + \delta_1 \sin^2 \theta_0\right)$$

$$\rho_0 = \rho_{op} \left(1 + a_2 \varepsilon + \delta_2 \sin^2 \theta_0\right)$$

$$B_0 = B_{op} \left(1 + a_3 \varepsilon + \delta_3 \sin^2 \theta_0\right)$$

The constants $\tilde{a}_i$ slightly alter the specific   value of the Parker boundary conditions but do not affect spherical symme-

Table 3a Predicted values at 1 a.u.[a]

θ,deg	Velocity, km/sec			Magnetic field, 10^{-5} gauss			Density particles/cm^3	Temperature, K
	V_r	V_θ	V_ϕ	B_r	B_θ/B_r	B_ϕ/B_r		
0	314	0.0	0.0	4.72	0.0	0.0	12.7	1.83×10^5
10	314	-0.64	0.18	4.70	-2.03×10^{-3}	-0.25	12.7	1.83×10^5
20	314	-1.20	0.36	4.67	-3.80×10^{-3}	-0.49	12.6	1.82×10^5
30	315	-1.61	0.54	4.61	-5.12×10^{-3}	-0.71	12.4	1.82×10^5
40	315	-1.83	0.72	4.55	-5.81×10^{-3}	-0.92	12.2	1.81×10^5
50	315	-1.83	0.89	4.48	-5.80×10^{-3}	-1.09	12.0	1.81×10^5
60	316	-1.61	1.04	4.41	-5.10×10^{-3}	-1.23	11.8	1.80×10^5
70	316	-1.20	1.16	4.36	-3.78×10^{-3}	-1.33	11.7	1.80×10^5
80	317	-0.64	1.24	4.32	-2.01×10^{-3}	-1.40	11.6	1.79×10^5
90	317	0.0	1.27	4.31	0.0	-1.42	11.6	1.79×10^5

[a]Constant coronal boundary conditions: $T_0 = 2 \times 10^6$ K, $N_0 = 2 \times 10^6$ particles/cm^3, and $B_0 = 1.0$ gauss.

Table 3b   Predicted values at 1 a.u.[a]

θ,deg	Velocity, km/sec			Magnetic field, 10^{-5} gauss			Density	
	V_r	V_θ	V_ϕ	B_r	B_θ/B_r	B_ϕ/B_r	Particles/cm^3	Temp. K
0	313	0.0	0.0	7.79	0.0	0.0	7.02	1.87×10^5
10	313	-2.62	0.17	7.74	-8.37×10^{-3}	-0.25	6.97	1.86×10^5
20	313	-4.92	0.43	7.59	-1.57×10^{-2}	-0.49	6.83	1.86×10^5
30	314	-6.63	0.83	7.37	-2.11×10^{-2}	-0.71	6.61	1.84×10^5
40	316	-7.54	1.39	7.09	-2.39×10^{-2}	-0.91	6.35	1.83×10^5
50	317	-7.54	2.06	6.80	-2.38×10^{-2}	-1.08	6.06	1.81×10^5
60	318	-6.65	2.76	6.53	-2.09×10^{-2}	-1.22	5.80	1.79×10^5
70	319	-4.92	3.37	6.30	-1.54×10^{-2}	-1.32	5.58	1.78×10^5
80	320	-2.62	3.80	6.16	-8.19×10^{-3}	-1.37	5.44	1.77×10^5
90	320	0.0	3.95	6.10	0.0	-1.39	5.40	1.77×10^5

[a]Constant coronal boundary conditions: $T_0 = 2 \times 10^6$ K, $N_0 = 1 \times 10^6$ particles/cm^3, and $B_0 = 1.5$ gauss.

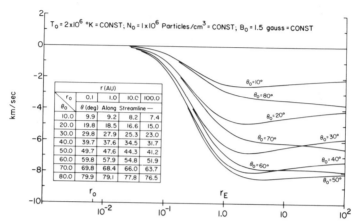

Fig. 5   Latitudinal flow velocity v_θ as a function of r along various streamlines for the coronal conditions used in Fig. 4b and Table 3b. The bending of each streamline away from the radial direction also is indicated.

try in the flow, since they do not appear in the radial differential equation. The constants δ_i represent the relative differences between the polar and equatorial values at the boundary r_0, or

$$[T_0(\pi/2) - T_0(0)]/T_0(0) = \delta_1\varepsilon + 0(\varepsilon^2)$$

with identical expressions between ρ_0 and δ_2, and B_0 and δ_3.

Table 4 is a comparison of the results of machine integration at 1 a.u. and 80°. The effect of a 5% positive or negative relative difference between the pole and equator is given for each variable separately. Note that $(\delta_1, \delta_2, \delta_3) = (0, 0, 0)$ corresponds to the solution for constant boundary conditions given in Table 3b and is included here for comparison. This table reveals that latitudinal flow at the orbit of Earth is most sensitive to latitudinal variations in the coronal temperature $(\delta_1, \delta_2, \delta_3) = \pm 0.05, 0, 0)$, and least sensitive to variations in the coronal magnetic field magnitude $(\delta_1, \delta_2, \delta_3) = (0, 0, \pm 0.05)$. A field stronger at the poles than at the equator r_0 also causes magnetic channeling of the flow toward the equator at distances near the corona. This is demonstrated in Fig. 6 by a positive latitudinal flow velocity near r_0. The dashed portion of the curve is an extension of the solution to inside the corona and is presented to demonstrate that ideally the model flow velocity approaches zero at the origin. This model is not valid inside the corona for several reasons, the most obvious being the rapid temperature increase between the solar surface and the corona. If the temperature

Table 4 Predicted values at Earth orbit with latitudinal dependence in coronal boundary conditions[a]

Boundary Conditions			Velocity, km/sec			Magnetic field, γ			Density	
T_0 δ_1	ρ_0 δ_2	B_0 δ_3	V_r	V_θ	V_ϕ	B_r	B_θ/B_r	B_ϕ/B_r	Particles/cm^3	Temp. K
+0.05	0	0	344	-1.00	1.07	4.21	-2.89×10^{-3}	-1.29	11.4	1.95×10^5
0	+0.05	0	317	-0.70	1.24	4.30	-2.19×10^{-3}	-1.40	11.7	1.79×10^5
0	0	+0.05	317	-0.64	1.29	4.47	-2.02×10^{-3}	-1.40	11.6	1.79×10^5
0	0	0	317	-0.64	1.24	4.32	-2.01×10^{-3}	-1.40	11.6	1.79×10^5
0	0	-0.05	316	-0.63	1.19	4.17	-2.00×10^{-3}	-1.40	11.6	1.79×10^5
0	-0.05	0	317	-0.58	1.24	4.34	-1.83×10^{-3}	-1.40	11.5	1.80×10^5
-0.05	0	0	289	-0.28	1.42	4.43	-0.96×10^{-3}	-1.53	11.8	1.64×10^5

[a]$r = 1$ a.u., $\theta = 80°$, $\varepsilon = 0.03$, $T_{po} = 2 \times 10^6$ K, $N_{po} = 2 \times 10^6$ particles/cm^3, and $B_{po} = 1$ gauss.

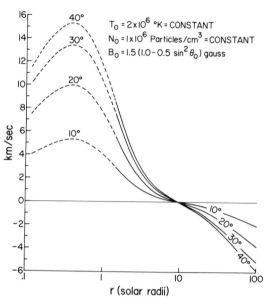

Fig. 6 Latitudinal flow velocity v_θ vs. r for a coronal field stronger at the poles than at the equator. The dashed sections of the curves represent the solution inside the corona but do not correspond to the actual behavior (see text). At radial distances greater than those shown, the curves are nearly identical to those in Fig. 4b corresponding to constant coronal boundary conditions.

is to increase with distance, than α must be reset to a value less than 1.0 in that region to employ the polytropic law. A proper treatment would require some knowledge of the heat sources. ces. The model also does not account for the density discontinuity at the solar surface. At larger distances, the curves shown in Fig. 6 approach the constant boundary condition curves shown in Fig. 3b, so that V_θ and the ratio B_θ/B_r at 1 a.u. are not affected significantly by the magnetic boundary variation.

In this discussion, we have not considered solutions that are not symmetric across the equatorial plane; however, such considerations are within the scope of the model. For example, a hot temperature band at a higher northern latitude would produce flow across the equator toward the southern hemisphere.

Discussion

In the foregoing, we have shown that the electromagnetic body forces in the solar wind will produce measurable effects on the solar wind flow. In all of the models treated here, the solar magnetic field and the solar wind do not vary. Yet the sun is never in a truly steady state. Consequently, these electric currents must vary as well, and their variations are probably responsible for certain of the waves and other disturbances recorded in the interplanetary magnetic field. How these currents grow and decay, what instabilities they are subject to, and what is their fate at great distances from the sun are questions that remain to be answered.

References

[1]Ferraro, V. C. A. and Bhatia, V. B., "Corotation and Solar Wind in the Solar Corona and Interplanetary Medium," Astrophysical Journal, Vol. 147, 1967, p. 220.

[2]Weber, E. J. and Davis, L., Jr., "The Angular Momentun of the Solar Wind," Astrophysical Journal, Vol. 148, 1967, p. 217.

[3]Modisette, J. L., "Solar Wind Induced Torque on the Sun," Journal of Geophysical Research, Vol. 72, 1967, p. 1521.

[4]Mestel, L., "Magnetic Braking by a Stellar Wind - I," Monthly Notices of the Royal Astronomical Society, Vol. 138, 1968, p. 359.

[5]Mestel, L., "Magnetic Braking by a Stellar Wind - II," Monthly Notices of the Royal Astronomical Society, Vol. 140, 1968, p. 177.

[6]Parker, E. N., "Dynamics of the Interplanetary Gas and Magnetic Fields," Astrophysical Journal, Vol. 128, 1958, p. 664.

[7]Parker, E. N., Interplanetary Dynamical Processes, Interscience, New York, 1964.

[8]Coleman, P. J., Jr., "Electric Currents in the Solar Wind," Journal of Geophysical Research, Vol. 80, 1975, p. 4719.

[9]Winge, C. R., Jr., "Latitude Effects in the Solar Wind," Ph.D. Dissertation, 1971, University of California, Los Angeles, Calif.

[10]Winge, C. R., Jr., and Coleman, P. J., Jr., "First Order Latitude Effects in the Solar Wind," Planetary and Space Science, Vol. 22, 1974, p. 439.

[11]Suess, S. T., "Three-Dimensional Solar Wind," Journal of Geophysical Research, Vol. 77, 1972, pp. 567-574.

[12]Suess, S. T., and Nerney, S. F., "Meridional Flow and the Validity of the Two-Dimensional Approximation in Stellar Wind Modeling, Astrophysical Journal, Vol. 184, 1973, pp. 17-25.

INTERACTION BETWEEN THE SOLAR WIND
AND THE INTERSTELLAR MEDIUM

Thomas E. Holzer*

National Center for Atmospheric Research,
Boulder, Colorado

Abstract

A review is given of the important physical processes
involved in the interaction of the solar wind with the interstellar medium. Four separate components of the interstellar
medium will be considered: 1) the interstellar neutral gas,
2) galactic cosmic rays, 3) the interstellar thermal plasma,
and 4) the galactic magnetic field. The neutral gas and the
cosmic rays exert a body force on the solar wind, tending to
decelerate continuously the supersonic flow, whereas the
thermal plasma and magnetic field exert a surface force on
the solar wind, tending to terminate the supersonic flow
abruptly through a shock transition. The individual effects
and the net effect of the four components are considered.

Introduction

There are four distinct components of the interstellar
medium that may have a significant effect on the solar wind
expansion: (1) the interstellar neutral gas; (2) galactic
cosmic rays; (3) the interstellar thermal plasma; and (4) the
galactic magnetic field. These four interstellar components
interact with the solar wind in two fundamentally different

Presented as Paper 73-548 at the AIAA/AGU Space Science
Conference on the Exploration of the Outer Solar System,
Denver, Colo., July 10-12, 1973. Part of this work was completed
while the author was a National Research Council Resident Re-
search Associate at the Aeronomy Laboratory of the Environmental
Research Laboratories of the National Oceanic and Atmospheric
Administration, Boulder, Colo. The National Center for
Atmospheric Research is sponsored by the National Science
Foundation.
*Member of Professional Staff, High Altitude Observatory.

ways. The neutral gas and cosmic rays penetrate deeply into
the interplanetary medium and exert a body force on the supersonic solar wind that tends to produce a continuous deceleration and heating of the solar wind. In contrast, the thermal
plasma and magnetic field exert a surface force at the outermost boundary of the heliosphere, tending to produce an
abrupt termination of supersonic solar wind flow through a
shock transition.

The entire subject of the interaction of the solar wind
with the interstellar medium has recently been discussed in
detail in an excellent review by Axford.[1] Rather than repeat
the discussions given by Axford[1], we shall consider the basic
physical processes involved in the interaction from a slightly
different point of view, giving primary emphasis to work
carried out subsequent to Axford's[1] review. The interested
reader is referred to Axford[1] for a detailed list of references
and for a more thorough discussion of work carried out prior
to mid-1971.

Deceleration of the Supersonic Solar Wind and the Transition to Subsonic Flow

The relatively large speed (~ 20 km sec^{-1}) of the sun
relative to the local standard of rest leads one to expect a
substantial relative motion between the sun and the local
interstellar medium. Such a relative motion results in the
penetration of the interstellar neutral gas deep into the inner
solar system.[2-5] For sufficiently large interstellar densities,
the presence of the interstellar neutral gas in interplanetary
space can have significant consequences for the solar wind[6,7]
expansion. Ogo 5 observations of the Ly-α sky background
indicate that the local interstellar atomic hydrogen density
is approximately 0.1 cm^{-3}, and perhaps as high[8] as 0.3 cm^{-3}.

Owing to charge exchange and photoionization, the solar
wind and the solar photon flux produce a cavity inside which
the interstellar atomic hydrogen density is severely attenu-
ated. The shape of the cavity and the sharpness of its
boundary depend on the relative motion between the sun and
the interstellar gas, the temperature of the interstellar gas,
the magnitude of solar radiation pressure, and asymmetries in
the solar wind and solar photon fluxes.[1,9-11] In general,
the cavity will have a minimum radius in the direction from
which the interstellar gas is flowing and will have a maximum
radius (as well as a more diffuse boundary) in the opposite

direction (i.e. downstream in the interstellar wind). For simplicity we shall restrict our attention to the upstream direction (i.e. the direction of minimum cavity radius), but our conclusions should also apply reasonably well to all other directions, including the downstream direction, since the Ogo-5 Ly-α sky background measurements indicate that the upstream and downstream density distributions are not drastically different,[12] but recent observations may change this conclusion.[36]

In order to understand the effects of the interstellar atomic hydrogen on the solar wind, we must consider the physical processes coupling the neutral gas to the solar wind plasma. As mentioned above, the important coupling processes are resonant charge exchange and photionization. In the resonant charge exchange process, protons with a mean flow speed and energy characteristic of the solar wind are lost, and protons with a mean flow speed and energy characteristic of the atomic hydrogen gas are produced; whereas photoionization produces protons with a mean flow speed and energy characteristic of the atomic hydrogen gas and electrons with a mean energy characteristic of the excess energy of the ionizing radiation. Evidently, both processes lead to the production of slow-moving protons, which are rapidly accelerated to high speeds by the magnetized solar wind. Momentum conservation tells us that each time the solar wind accelerates one of these newly-produced protons, the solar wind speed must be reduced slightly. Thus, the slow-moving protons produced through charge exchange and photoionization exert a friction-like force that tends to decelerate the supersonic solar wind (associated heating effects are discussed in section 3).

Most solar wind theories have considered no retarding body force other than solar gravity, which rapidly becomes negligible beyond several solar radii. Consequently, the retarding body force associated with the friction-like interaction between the interestellar neutral gas and the solar wind can, in principle, lead to a family of solar wind solutions fundamentally different from that obtained by neglecting the interstellar neutral gas. Such families have been computed[13-15] for an unmagnetized solar wind and an atomic hydrogen gas with density n_H = const in $r \geq r_o$ and n_H = 0 in $r < r_o$. The families of solutions shown in Figures 1a and 1b are taken from Holzer's[15] results and are only valid in $r \geq r_o \approx 5$ AU. It is seen that the interstellar neutral gas leads to a new critical point in the solar wind solutions, which takes the form of a node (Fig. 1a) or a focus (Fig. 1b), unlike the saddle-point form of the familiar critical point near the sun.[13] Clearly the node provides the possibility for a smooth transition from supersonic to subsonic solar wind flow, whereas the focus requires a shock transition, just as did previous solar wind theories.[15] From computations including the interplanetary magnetic field[15] it is found that for realistic solar wind and interstellar gas parameters the critical point generally has the form of a node (Fig. 1a), so that a shock-free transition from supersonic to subsonic solar wind flow can, in principle, exist.[16]

However, in a realistic solar-wind model, the requirements for the existence of such a shock-free transition are rather severe. These requirements are: (1) for the given parameters of the system, one or more shock-free solutions must exist, (2) the supersonic solution that satisfies the boundary conditions at the inner edge of the system must correspond to one of the available shock-free solutions; and (3) one of the possible subsonic continuations of the supersonic shock-free solution must satisfy the pressure balance boundary condition at the outer edge of the system. If any one of these requirements is not satisfied, it is necessary to insert a shock transition at a suitable location.[17] In practice, requirements 1 and 2 will be satisfied, but, as is discussed below, requirement 3 appears not to be satisfied unless the local interstellar atomic hydrogen density is somewhat higher than is currently thought.

In considering requirement 3 and the associated externally imposed boundary conditions, we are brought to the problem of the interaction between the solar wind and the magnetized interstellar plasma. Since the solar wind and the interstellar thermal plasma are both highly conducting and magnetized, there will be a tendency for the plasmas not to interpenetrate. Hence, we shall assume that there is a relatively distinct boundary separating the solar wind from the interstellar thermal plasma and magnetic field, and we shall call this boundary the heliopause. In a steady state, there should be a pressure balance across the heliopause, so that the total pressure of the interstellar plasma and magnetic field provides a boundary condition on the solar wind expansion. In fact, it is this boundary condition that ultimately determines whether the solar wind is a supersonic or a subsonic expansion.[17] As it happens, the interstellar pressure is much too small to inhibit the initial transition from subsonic to supersonic flow (at $r = r_c$), but the fact that the pressure

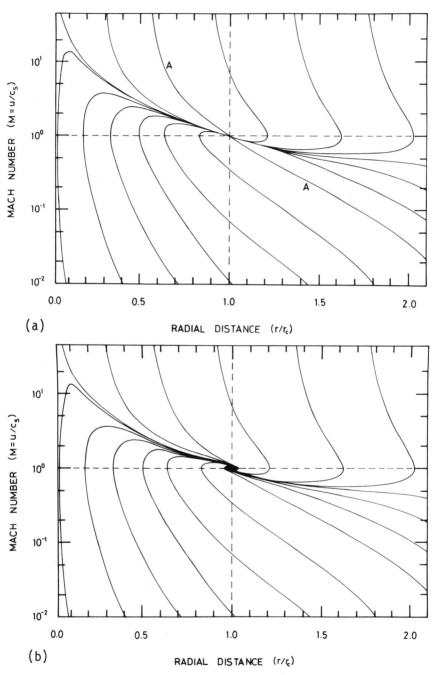

(a)

RADIAL DISTANCE (r/r_c)

(b)

RADIAL DISTANCE (r/r_c)

Fig. 1   Families of solutions to the solar wind equations
        showing two types of critical points --(a) node,
        (b) focus   (after Holzer[15]; see also Wallis[13,14]).

is non-zero requires that there exist (much farther from the
sun, at $r = r_s \gg r_c$) a transition from supersonic to subsonic
solar wind flow. In the absence of any friction-like deceleration (such as that associated with the interstellar neutral
gas), this transition will necessarily be a shock.[17-22] It
is clear why a transition to subsonic flow is required, when
one realizes that along a supersonic solution the solar wind
ram pressure decreases monotonically with increasing r, so
that eventually the total solar wind pressure falls below the
interstellar pressure, unless the flow becomes subsonic.

Let us now see what we can learn about the nature,
the location, and the geometrical shape of the heliopause and
of the transition from supersonic to subsonic flow. (Both the
transition and the heliopause should form closed, or nearly
closed, surfaces in three-dimensional configuration space.)
First, we shall consider the point on the transition surface
$(r_s(\theta,\phi) = r_{so})$ and the point on the heliopause $(r_H(\theta,\phi) = r_{Ho})$
which are nearest the sun. These two points should lie very
nearly along the same radius vector (θ_o,ϕ_o), which is
determined by the point just outside the heliopause where the
total interstellar pressure takes on its maximum value.
This radius vector traces out (from the sun to the heliopause)
a solar wind stagnation flow line. Hence, the total solar wind
pressure at the point (r_{Ho},θ_o,ϕ_o) will be the sum of the
thermal and magnetic pressures (i.e. $p_{Ho} + B_{Ho}^2/8\pi$). The
maximum interstellar pressure is given by $\rho_i u_i^2 + p_i + \alpha B_i^2/8\pi$,
where ρ_i, u_i, p_i, and B_i are the interstellar thermal plasma
mass density, flow speed (relative to the sun), pressure, and
magnetic field strength, all measured well away from the
heliopause; the factor takes account of the effect of solar
wind distortion of the interstellar magnetic field.[20,1] (If
$\rho_i u_i^2 \gg B_i^2/8\pi$, then $\alpha \approx 1$, but if $B_i^2/8\pi > \rho_i u_i^2$, then $\alpha > 1$.)
Thus, at the stagnation point on the heliopause (r_{Ho},θ_o,ϕ_o),
the pressure balance condition becomes

$$p_{Ho} + B_{Ho}^2/8\pi = \rho_i u_i^2 + p_i + \alpha B_i^2/8\pi \qquad (1)$$

If, for the moment, we neglect the interstellar neutral
gas and the interplanetary magnetic field, then the transition
from supersonic to subsonic flow must be a shock, and it is

readily shown[20,1] that the solar wind flow in the subsonic region downstream of the shock is essentially incompressible. Hence, we can equate the solar wind ram pressure just upstream of the shock ($\rho_{so} u_{so}^2$) to the stagnation pressure (p_{Ho}), and since the ram pressure of the supersonic solar wind varies nearly as r^{-2} (i.e. for n_H 0), we have

$$r_{so} \approx \left[\rho_{Eo} \, u_{Eo}^2 \, r_E^2 / (\rho_i \, u_i^2 + p_i + \alpha \, B_i^2/8\pi)\right]^{\frac{1}{2}} \qquad (2)$$

where the subscript 'E' refers to the orbit of the earth.

For reasonable solar wind and interstellar parameters,[1] (2) yields a minimum shock distance of $r_{so} \approx 100$ AU. With this value in hand, we can return to consideration of the third requirement for a shock-free transition, before completing our discussion of the shape of the heliopause and of the supersonic-subsonic transition. Figure 2 shows the rate of slowing of the supersonic solar wind for several interstellar neutral hydrogen densities. Evidently, for all densities $n_H \geq 0.1$ cm^{-3}, there is a significant slowing of the solar wind before the nominal minimum shock distance of 100 AU

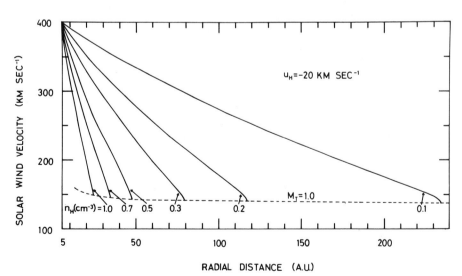

Fig. 2   Radial profiles of solar wind speed for various values of n_H. The dashed line represents the locus of sonic points   (after Holzer[15]).

is reached (for $n_H = 0$, the solar wind speed remains nearly
constant out to the shock). This slowing implies a more rapid
decrease in ram pressure than the r^{-2} dependence assumed in
deriving (2) (viz. $\rho u^2 \sim u/r^2$). Hence for finite n_H, the
minimum shock distance should actually be less than 100 AU.
We see from Figure 2 that for $n_H = 0.1$ cm^{-3}, r_{so} (as given
by (2)) is reduced by about 15%, whereas for $n_H = 0.2$ cm^{-3}
and $n_H = 0.3$ cm^{-3} the reductions are about 20% and 30%.

Another factor that reduces r_{so} still further is the existence
of an interplanetary magnetic field, for in the subsonic
region the magnetic field rapidly takes control of the flow
and leads to a compressible medium. As can be seen in
Figure 3, the decrease in pressure between the shock and
the heliopause depends on the thickness of the subsonic
region. For $r_{Ho}/r_{so} \approx 3$ there is a 40% pressure reduction and
a consequent 20% reduction in r_{so}. (Note that for large
values of r_{Ho}/r_{so} the curves of Figure 3 undoubtedly decrease
too steeply, since the effects of field-line reconnection at
sector boundaries have not been included.) A further effect
tending to produce an inward-directed pressure gradient in the
subsonic region is the tendency for flow lines to diverge
(more rapidly than a radial divergence) in the vicinity of a
stagnation line.

The net result of all the effects discussed above is
to reduce significantly the minimum distance to the shock
transition. Apparently if $n_H = 0.1$ cm^{-3}, then $r_{so} \lesssim 60$ AU,
whereas if $n_H = 0.3$ cm^{-3}, then $r_{so} \lesssim 45$ AU. Returning our
attention to Figure 2, it is clear that along the line (θ_0, ϕ_0)
the sonic point for the shock-free transition is always at
much larger radial distances than is the expected distance
to the shock transition. Hence, at least in the direction
(θ_0, ϕ_0), we expect the transition from supersonic to subsonic
solar wind flow to involve a shock discontinuity. If the
speed of the interstellar medium relative to the sun is significant ($u_i \gtrsim 10$ km sec^{-1}), it is reasonable to assume that
the heliopause boundary has a shape similar to that of the
terrestrial magnetopause. Then the pressure boundary condition in the tail of the heliosphere should be determined

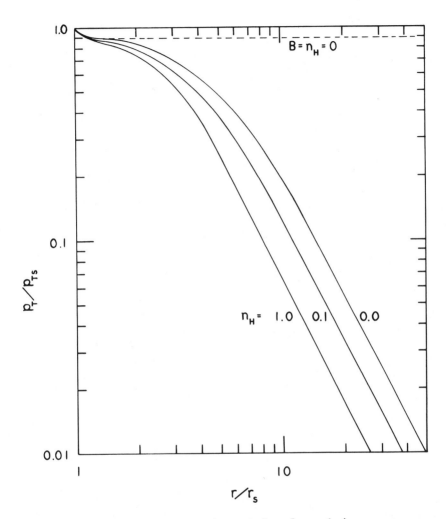

Fig. 3 Radial profiles of the total solar wind pressure
(p_T) in the postshock region. The subscript 's'
refers to the shock transition (after Holzer[15]).

largely by the interstellar magnetic pressure and thus should
be no less than half of the maximum interstellar pressure
applied at the stagnation point on the heliopause nose. It
follows that the shock distance in the tail should be no more
than 25% larger than the minimum shock distance, and therefore
that the transition from supersonic to subsonic flow should be
a shock discontinuity in all directions, provided $n_H \lesssim 0.3$ cm^{-3}.

So far we have neglected the effects of galactic cosmic
rays and of a non-steady solar wind on the shape and location
of the shock transition. The problem of solar wind modification
by galactic cosmic rays has been considered by several
authors.[23-26] Although qualitatively the cosmic rays affect
the solar wind in much the same way as does the neutral interstellar gas, it appears that the magnitude of this effect is
quite small in comparison with that of a neutral interstellar
gas of density $n_H \gtrsim 0.1$ cm^{-3}. On the other hand, the solar
wind stream structure[27] may have a somewhat more noticeable
(transient) effect on the shock location. If the large scale
structures observed at 1 AU are not damped out by turbulent
dissipation,[28] they may persist to well beyond 10 AU[29] and
perhaps all the way out to the shock transition. If the
structures do persist, then we might expect a quasi-periodic
inward and outward motion of the shock on a time scale of
several days. The total shock displacement probably should
not be more than a few AU, and the mean shock position
should be displaced slightly outward from the expected
steady position for the mean solar wind energy density.
Evidently such transient effects, though interesting in
themselves, do not seriously modify the basic shock
morphology discussed above.

Heating of the Supersonic Solar Wind and Cooling of the Subsonic Solar Wind

As was mentioned in the preceding section, the interplanetary atomic hydrogen gas is coupled to the solar wind
through the processes of resonant charge exchange and photoionization, and this coupling leads to a friction-like interaction that tends to slow and heat the supersonic solar wind.
The slowing process was discussed in section 2, and we shall
now go on to consider the heating process, following the
basic approach of Holzer and Leer.[30] In this discussion we
shall assume, for simplicity, that each proton formed from
the ionization of an interplanetary hydrogen atom is produced
initially at rest in a heliocentric reference frame. The
neglect of the atomic hydrogen motion relative to the sun is
reasonable in a qualitative discussion, since this relative
motion is quite small in comparison with the solar wind speed
in the heliocentric rest frame.

A newly-produced stationary proton may interact with
the solar wind in a number of ways, and may eventually become

indistinguishable from other solar wind protons. The modes
of interaction include acceleration by the interplanetary
magnetic field (IMF), Coulomb collisions, and wave-particle
interactions. However, regardless of the mode(s) of interaction, the newly-produced proton cannot change the total
energy density or momentum density of the solar wind (in our
heliocentric rest frame), since the proton is produced at
rest. (Of course, in the charge-exchange interaction, a solar
wind proton is lost, and this loss decreases both the total
energy density and the momentum density of the solar wind.)
Because of the large interaction speeds, Coulomb collisions
cannot be important, but the IMF and wave-particle interactions should, in general, be important. In the presence of
an IMF, the newly produced proton will be accelerated in-
stantaneously so that it is travelling with an average
velocity (normal to the local magnetic field) of magnitude
$V \sin\psi$ (V is the solar wind speed and ψ is the angle between
the local IMF and the heliocentric radius vector). In
addition the proton will be executing a circular motion
about the field lines characterized by the same speed and
thus will have gained a total energy of $m_p V^2 \sin^2 \psi$, where
m_p is the proton mass. From an examination of the conservation equations,[30] it is clear that virtually all (viz. a
fraction $1-M^{-2}$, where M is the solar wind Mach number) the
energy gained by the newly-formed proton is derived from
solar wind bulk flow (as opposed to thermal) energy. If
wave-particle interactions lead to a randomization of the
proton's motion, so as to make it indistinguishable from
solar wind protons,[30] then the conservation laws tell us
that the total energy of the proton <u>in the solar wind rest</u>
<u>frame</u> $\left[\text{i.e.} \approx \frac{1}{2} m_p V^2 \sin^2 \psi \text{ (circular motion about field}\right.$
lines$) + \frac{1}{2} m_p V^2 \cos^2 \psi$ (motion along field lines)$\Big]$ must go

into thermal energy of the solar wind protons. Consequently,
if a newly-produced proton is thermalized (i.e. becomes
indistinguishable from solar wind protons), then the net
result of the thermalization interaction is a transformation
of solar wind bulk flow energy $(\approx \frac{1}{2} m_p V^2)$ into solar wind
thermal energy, and this transformation leads to a net heating
and slowing of the supersonic solar wind. However, since
the solar wind is highly supersonic, the fractional decrease
of flow energy is much smaller than the fractional increase
of thermal energy, so the temperature increase must be much
more significant than the velocity decrease.

The relatively large heating and small deceleration effects of the neutral gas on the solar wind are evident in Figure 4, which is taken from Holzer and Leer.[30] This figure shows radial profiles between 1 AU and 10 AU of the solar wind bulk flow speed (V), electron temperature (T_e), and proton temperature (T_p) for several different interstellar atomic hydrogen densities ($n_H = n_{H\infty} \exp(-A/r)$; A = 4 AU). Temperature profiles for two types of electron-proton thermal coupling are shown. In one case, it is assumed that the electron and proton gases are strongly coupled if the proton temperature exceeds (even slightly) the electron temperature, and in this case T_e and T_p exhibit the same profile beyond the point where $T_p(r)$ intersects $T_e(r)$. In the second case, it is assumed that only Coulomb collisions couple the electron and proton gases, and since this coupling is very weak in the region of interest, the ratio T_p/T_e exceeds unity beyond 4-6 AU. For $n_{H\infty} = 0.0$ (i.e. in the absence of interplanetary atomic hydrogen), the profiles of V, T_p, and T_e are just those that would be expected in a free solar wind expansion: the flow speed monotonically increases with increasing radial distance, whereas the electron and proton temperatures monotonically decrease with increasing radial distance. However, in the presence of an interplanetary atomic hydrogen gas ($0.1 \text{ cm}^{-3} \leq n_{H\infty} \leq 0.3 \text{ cm}^{-3}$), the flow speed reaches a maximum in 3 AU < r < 6 AU, thereafter monotonically decreasing with increasing radial distance; the proton temperature reaches a minimum in 2 AU $\leq$ r $\leq$ 4 AU, thereafter monotonically increasing with increasing radial distance; and the electron temperature either decreases normally (weak coupling) or reaches a minimum in 4 AU $\leq$ r $\leq$ 6 AU, thereafter following the proton temperature. Of course, some intermediate behaviour of the electron temperature is also possible, and perhaps is most likely. (We note that solar wind α-particles are likely to exhibit a radial temperature profile similar to that of protons.) Strong coupling temperature profiles are extended to larger radial distances in Figure 5, where it is assumed that n_H = constant beyond 5 AU. These profiles of thermal speed $\left[= (5k(T_e + T_p)/3m_p)^{\frac{1}{2}} \right]$ correspond to the solar wind velocity profiles shown in Figure 2.

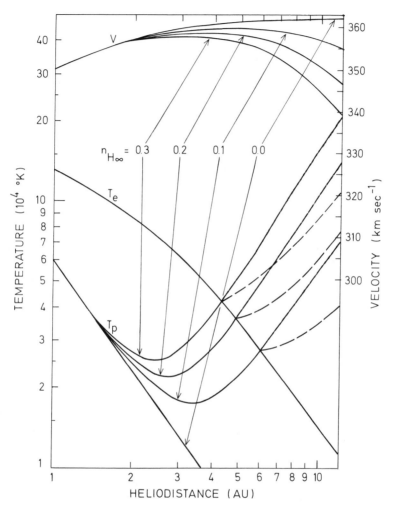

Fig. 4   Radial profiles of V, T_e, and T_p for several
values of $n_{H\infty}$. The solid portions of the T_e and T_p
curves correspond to weak coupling (Coulomb collisions)
between protons and electrons, whereas the dashed
portions of the T_e and T_p curves, which appear only
beyond the point where $T_p = T_e$, correspond to strong
coupling between protons and electrons (maintaining
$T_e = T_p$). Only one weak coupling T_e profile is
shown, since T_e is only very weakly dependent on $n_{H\infty}$
when electron heating by waves is neglected   (after
Holzer and Leer[30]).

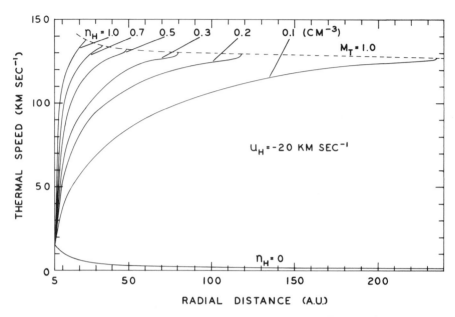

Fig. 5 Radial profiles of solar wind thermal speed for
        various values of n_H. The dashed line shows the
        locus of sonic points   (after Holzer[15]).

An interesting consequence of the heating of the solar
wind relates to the β parameter of the plasma. Figure 6 shows
radial profiles of the ratio c_A^2/c_s^2 ($=B^2/4\pi\gamma p \approx \beta^{-1}$), corresponding to the profiles of solar wind velocity and thermal
speed shown in Figures 2 and 5. Evidently, in the absence
of interplanetary hydrogen the plasma β, which is of the
order of 1 at 1 AU, decreases quite rapidly with increasing
radial distance owing to adiabatic cooling of the plasma.
However, the heating of the solar wind associated with
interplanetary hydrogen causes β to remain large (1 < β < 10)
beyond 1 AU. Axford[1] has discussed the possibility that
field-line reconnection does not take place at sector
boundaries in r ≲ 1 AU because β (≳ 1) is too large. If
this is the case, one might expect reconnection to become
significant beyond 1 AU in the absence of a neutral gas. Hence
it is possible that interstellar hydrogen could play a significant role in inhibiting reconnection at sector boundaries
in the supersonic solar wind.

     In the postshock subsonic solar wind the primary effect of
of an atomic hydrogen gas is a contribution to the cooling of
the hot shocked plasma.[31] However, even in the absence of a

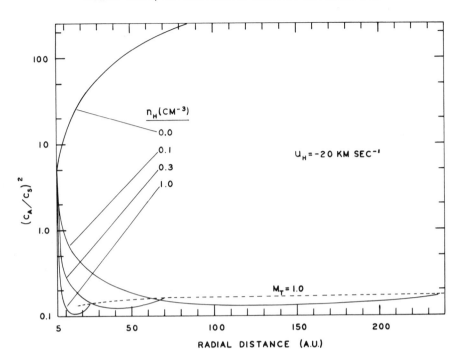

Fig. 6   Radial profiles of the inverse plasma β for various
values of n_H. The dashed line represents the locus
of sonic points   (after Holzer[15]).

neutral gas, the presence of an IMF will lead to a significant
cooling of the plasma. Of course, in the absence of both a
neutral gas and a magnetic field, there is only a slight
cooling immediately behind the shock, followed by a nearly
isothermal subsonic expansion (cf. also section 2). The
cooling effects of the neutral gas and the magnetic field may
be compensated to some extent by heating associated with field-
line reconnection, if this process becomes important in the
subsonic region (see below). One consequence of the cooling
process is the production of a population of hot hydrogen
atoms.[31-34,15] These hot hydrogen atoms can penetrate into
the region of supersonic solar-wind flow and potentially form
an important component of the interplanetary neutral gas. By
making use of a detailed model of the postshock region,
Holzer[15], employing a method similar to that of Hundhausen,[34]
has calculated the density of hot neutrals in the supersonic
region for various shock distances. The results of this
calculation indicate that unless the shock transition is
located in the inner solar system (unlikely in view of the

discussion in section 2), the hot atoms represent only a minor component of the interplanetary hydrogen gas.

In addition to the production of hot neutrals, the cooling of the hot post-shock solar-wind plasma is associated with the introduction of compressibility into the postshock flow and a probable enhancement of field-line reconnection at existing magnetic neutral surfaces (e.g. sector boundaries). The effect of compressibility on the location of the shock transition was discussed in section 2, and we shall now go on to discuss the possibility that field-line reconnection becomes an important process in the region of subsonic flow.

The β of a highly-conducting subsonic plasma ($\beta \approx c_s^2/c_A^2$) gives a measure of the degree of control that the magnetic field has over the dynamical behaviour of the plasma. In a high-β plasma ($c_A^2/c_s^2 \ll 1$) the dynamical behaviour of the plasma governs the magnetic field behaviour, whereas in a low-β plasma ($c_A^2/c_s^2 \gg 1$) the magnetic field largely controls the plasma dynamics. Consequently, in a high-β plasma ($c_A^2/c_s^2 \ll 1$) the existence of a magnetic neutral sheet does not guarantee that there will be significant field-line reconnection, whereas in a low-β plasma ($c_A^2/c_s^2 \gg 1$) we can expect rapid reconnection at magnetic neutral sheets. Field-line reconnection acts to heat the plasma and to establish a plasma pressure gradient that tends to inhibit the reconnection process. Consequently, reconnection should act to increase the β of a low-β plasma until β becomes of order 1.

From Figure 7 we see that $\beta \gtrsim 1$ in $r_s < r < 3r_s$ and that $\beta \lesssim 1$ in $r > 3r_s$. Thus if the heliopause is located beyond $3r_s$ (cf. section 2), field-line reconnection may be significant in the outer part of the subsonic region. If this is the case, we should expect that in $r \gtrsim 3r_s$, $\beta \approx c_s^2/c_A^2 \approx 1$. Of course, a constant β in this region would lead to a decrease in the plasma compressibility. Even if there is no significant field-line reconnection within the heliosphere, we should expect the interplanetary magnetic field to become connected to the interstellar magnetic field at the heliopause,[1] and it is this connection that should limit the length of an ordered heliospheric tail.

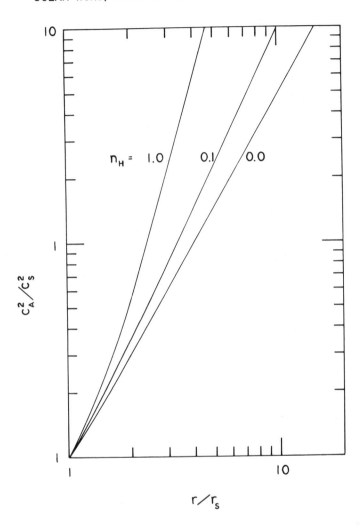

Fig. 7   Radial profiles of the inverse plasma β in the postshock region. The shock distance is r_s.

Summary

   We have seen that the interstellar medium and the solar
wind exhibit two fundamentally different types of interaction.
The magnetized interstellar plasma exerts both a normal and a
tangential stress at the surface bounding the solar wind (i.e.
at the heliopause). The normal stress causes the solar wind to
undergo a transition from supersonic to subsonic flow, and the
tangential stress turns the subsonic solar wind flow, leading
to a heliospheric cavity similar in shape to the terrestrial

magnetosphere. The interstellar neutral gas penetrates into the inner solar system and exerts a friction-like force on the solar wind that tends to slow and heat the supersonic solar wind. For a sufficiently strong neutral gas interaction, a shock-free supersonic-subsonic solar wind transition is possible, but it appears that the interstellar atomic hydrogen density is so low that a shock transition must exist.

The manifestation of the interaction between the solar wind and the interstellar medium that is most likely to be observable by a spacecraft travelling into the outer solar system is the heating of the supersonic solar wind owing to the penetration of interstellar hydrogen into interplanetary space. Of course, this effect is likely to be obscured (at least partially) by the solar wind stream structure and any associated dissipative effects, so that spacecraft observers must be very careful in interpreting their data in this regard. A satellite that travels all the way to 50 AU before dying may have a chance of detecting the shock transition from supersonic to subsonic solar wind flow. Finally, we note that the penetration of the interstellar gas into the inner solar system (viz. interstellar He) can lead to observable changes in the solar wind ionization state.[35]

References

[1] Axford, W. I., "The Interaction of the Solar Wind with the Interstellar Medium," in Solar Wind, edited by C. P. Sonett, P. J. Coleman, Jr., and J. M. Wilcox, NASA SP-308, 1972, p. 609.

[2] Blum, P. W. and Fahr, H. J., "Solar Wind Tail and the Anisotropic Production of Fast Hydrogen Atoms," Nature, Vol. 223, 1969, p. 936.

[3] Blum, P. W. and Fahr, H. J., "Interaction between Interstellar Hydrogen and the Solar Wind," Astron. Astrophys., Vol. 4, 1970, p. 280.

[4] Holzer, T. E., "Stellar Winds and Related Flows," Ph.D. Thesis, Univ. of Calif., San Diego, 1970.

[5] Holzer, T. E. and Axford, W. I., "The Interaction between Interstellar Helium and the Solar Wind," J. Geophys. Res., Vol. 76, 1971, p. 6995.

[6]Thomas, G. E. and Krassa, R. F., "Ogo-5 Measurements of the Lyman Alpha Sky Background," _Astron. Astrophys._, Vol. 11, 1971, p. 218.

[7]Bertaux, J. L. and Blamont, J. E., "Evidence for a Source of an Extraterrestrial Hydrogen Lyman-Alpha Emission: the Interstellar Wind," _Astron. Astrophys._, Vol. 11, 1971, p. 200.

[8]Wallis, M. K., "Local Hydrogen Gas and the Background Lyman-Alpha Pattern," submitted to _Mon. Not. Roy. Astron. Soc._, 1973.

[9]Thomas, G. E., "Properties of Nearby Interstellar Hydrogen Deduced from Lyman α Sky Background Measurements," in _Solar Wind_, edited by C. P. Sonett, P. J. Coleman, Jr., and J. M. Wilcox, NASA SP-308, 1972, p. 668.

[10]Thomas, G. E., "The Neutral Interplanetary Medium," AIAA Paper No. 73-547, July 1973, Denver, Colo.

[11]Johnson, H. E., "Backscatter of Solar Resonance Radiation-I," _Planet. Space Sci._, Vol. 20, 1972, p. 829.

[12]Holzer, T. E., "Interpretation of Ly-α Sky Background Observations," _EOS_, November, 1973.

[13]Wallis, M. K., "Transonic Deceleration of the Solar Wind via Planetary, Cometary, or Interstellar Gas," Report TRITA-EPP-71-01, Royal Institute of Technology, Stockholm, 1971, 24 pp.

[14]Wallis, M. K., "Shock-Free Deceleration of the Solar Wind," _Nature_, Vol. 233, 1971, p. 23.

[15]Holzer, T. E., "Interaction of the Solar Wind with the Neutral Component of the Interstellar Gas," _J. Geophys. Res._, Vol. 77, 1972, p. 5407.

[16]Parker, E. N., "Theoretical Studies of the Solar Wind Phenomenon," _Space Sci. Rev._, Vol. 9, 1969, p. 325.

[17]Holzer, T. E. and Axford, W. I., "The Theory of Stellar Winds and Related Flows," <u>Ann. Rev. Astron. Astrophys.</u>, Vol. 8, 1970, p. 31.

[18]Clauser, F., Johns Hopkins Univ. Lab. Rept. AFOSRTN 60-1386, 1960.

[19]Weymann, R., "Coronal Evaporation as a Possible Mechanism for Mass Loss in Red Giants," <u>Astrophys. J.</u>, Vol. 132, 1960, p. 380.

[20]Parker, E. N., "The Stellar Wind Regions," <u>Astrophys. J.</u>, Vol. 134, 1961, p. 20.

[21]Parker, E. N., <u>Interplanetary Dynamical Processes</u>, Interscience, New York, 1963, 272 pp.

[22]McCrea, W. H., "Shock Waves in Steady Radial Motion Under Gravity," <u>Astrophys. J.</u>, Vol. 124, 1956, p. 461.

[23]Axford, W. I. and Newman, R. C., "The Effect of Cosmic Ray Friction on the Solar Wind," <u>Proc. 9th Int. Conf. Cosmic Rays, Phys. Soc.</u>, London, 1965, p. 173.

[24]Sousk, S. F. and Lenchek, A. M., "The Effect of Galactic Cosmic Rays upon the Dynamics of the Solar Wind," <u>Astrophys. J.</u>, Vol. 158, 1969, p. 781.

[25]Wallis, M. K., "Interaction between the Interstellar Medium an and the Solar Wind," <u>Astrophys. Space Sci.</u>, Vol. 20, 1973, p. 3.

[26]Suess, S. T., Fisk, L. A., and Holzer, T. E., "Galactic Cosmic Ray Modulation in the Outer Solar System: Solar Wind Consequences," <u>NOAA Tech. Rept. ERL 285-SEL 26</u>, U. S. Dept. of Commerce, 13 pp., November, 1973.

[27]Hundhausen, A. J., <u>Coronal Expansion and Solar Wind</u>, Springer-Verlag, New York, 1972, 238 pp.

[28]Jokipii, J. R. and Davis, L., Jr., "Long-Wavelength Turbulence and the Heating of the Solar Wind," <u>Astrophys. J.</u>, Vol. 156, 1969, p. 1101.

[29]Hundhausen, A. J., "Evolution of Large-Scale Solar Wind Structures Beyond 1 AU," J. Geophys. Res., Vol. 78, 1973, p. 2035.

[30]Holzer, T. E. and Leer, E., "Solar Wind Heating beyond 1 AU," Astrophys. Space Sci., Vol. 23, 1973, p. 515.

[31]Axford, W. I., Dessler, A. J., and Gotlieb, B., "Termination of the Solar Wind and Solar Magnetic Field," Astrophys. J., Vol. 137, 1963, p. 1268.

[32]Patterson, T. N. L., Johnson, F. S., and Hanson, W. B., "The Distripution of Interplanetary Hydrogen," Planet. Space Sci., Vol. 11, 1963, p. 767.

[33]Dessler, A. J., "Solar Wind and Interplanetary Magnetic Field," Rev. Geophys., Vol. 5, 1967, p. 1.

[34]Hundhausen, A. J., "Interplanetary Neutral Hydrogen and the Radius of the Heliosphere," Planet. Space Sci., Vol. 16, 1968, p. 783.

[35]Holzer, T. E. and Axford, W. I., "Solar Wind Ion Composition," J. Geophys. Res., Vol. 75, 1970, p. 6354.

[36]Bohlin, R. C., "Mariner 9 Ultraviolet Spectrometer Experiment: Measurements of the Lyman-Alpha Sky Background," Astron. Astrophys., Vol. 28, Nov. 1973, pp. 323-326.

SOLAR WIND DISTURBANCES CAUSED BY
PLANETS AND SOLAR FLARES

Murray Dryer[*]

National Oceanic and Atmospheric Administration,
Boulder, Colorado

Abstract

The sun acts as a gaseous source for a "wind tunnel" on
the scale of the solar system itself. Both steady and un-
steady fluid phenomena within this plasma physics laboratory
have been reliably established for the solar wind's interaction with the Earth's magnetosphere and for solar flare-generated interplanetary shock waves. Armed with this background
and some observations, speculations are made regarding shock
propagation beyond one astronomical unit and solar wind interaction with Jupiter, Saturn, Uranus, Neptune and Pluto. Continuum MHD physics is used for this purpose because of its
success in the earlier explorations. For example, available
observational data at Jupiter are compared with the continuum
theory. The time-dependent studies of interplanetary disturbances look very promising but require additional comparisons
of observations and theory. Some discussion, then, is given
to some aspects of shocks (multiple, forward, reverse, etc.),
their generation at the sun, and their propagation through the
interplanetary medium.

Presented as Paper 73-561 at the AIAA/AGU Space Science
Conference: Exploration of the Outer Solar System, Denver,
Colorado, July 10-12, 1973. I wish to thank Drs. J. H. Wolfe,
H. R. Collard, and J. D. Mihalov for permission to discuss
their Pioneer 9 and 10 data prior to publication; Dr. P. A.
Penzo, for the use of the Grand Tour trajectories; and Drs. R.
F. Donnelly and D. S. Intriligator for suggestions during the
preparation of this paper.
*Senior Scientist, Space Environment Laboratory, Environ-
mental Research Laboratories.

Introduction

The outward flow of solar plasma (protons and electrons in approximately equal densities, alpha particles in an amount of ~ 4 percent of the protons, and minor traces of heavier ionized elements) represents a small fraction of the solar energy output. Thermonuclear fusion within the solar core converts 4.4 metric tons of mass each second into a power output of 4×10^{33} erg/sec. This energy leaks, via radiative transfer, into a convective layer which carries it to the photosphere, the visible solar surface. There, it escapes primarily as optical and infrared radiation as an energy flux of approximately 6×10^{10} erg/cm^2 sec. This primary energy flux is practically unaffected by solar activity (Evans[1]). A smaller flux of about 10^6 erg/cm^2 sec, composed of EUV, X-ray, radio radiation, and particles (mostly the solar wind) also escapes to space. The solar wind initiates its expansion at the coronal base by slowly (~ 3.5 km/sec) carrying away a few percent of this latter flux, primarily in the thermal state which—by the time it expands to sonic velocities at about 6 solar radii (the solar "throat" of a Venturi nozzle)—represents the source for an essentially spherically-symmetric supersonic "wind tunnel". The energy flux convected to Earth, under quiet conditions, amounts to about 0.2 erg/cm^2 sec.

It is of interest to examine briefly the comparative effects at Earth when: (1) the X-ray and EUV radiation reach the atmosphere, and (2) the solar wind reaches the terrestrial magnetic field. The energy contained in each of the principal line emissions and continua in the (average) solar spectrum within the wavelength range from 140 to 1340 Å (Allen[2]) amounts to ~ 0.02 to about 0.3 erg/cm^2 sec (at the top of the atmosphere) with a narrow maximum (of about 5 erg/cm^2-sec) at the extremely intense HI Lyman-α line at 1216 Å. As summarized by Reid[3], this EUV and X-radiation (the latter, from about 1 to 120 Å) is responsible for ionization of the atmospheric neutral constituents, thereby producing the D, E, and F regions of the ionosphere (from about 80 to 300 km). During solar flares, the Lyman-α energy flux has little variability; i.e., it increases only by a factor of 2 or less for a few minutes; whereas the 2-12 Å X-ray flux can increase by several orders of magnitude, i.e., from as low as 0.4×10^{-3} to 735×10^{-3} erg/cm^2-sec (Reid[3]). In any case, the solar flare effects are confined mainly to the ionosphere by these inputs. On the other hand, the quiet solar wind energy flux is deflected by the terrestrial magnetic field which is deformed into a comet-like configuration called the "magnetosphere". When solar flares occur, however, a sizeable fraction (about half) of the total flare energy is converted into mechanical (kinetic and thermal) energy and is added to the solar wind by means of an interplanetary shock wave. Table 1 shows estimates of the energy released in both large and small flares (Pintér[4]). It should be noted that the energy listed under the last item refers to that contained in the disturbed solar wind behind the shock wave in excess of that which this flux contained in the steady state prior to the flare.

A major goal of exploration in the outer solar system is the study of the interaction of the quiet (or average) solar wind with the planets beyond Earth. Additionally, the response of the solar wind at large distances to solar flares is an equally important goal; this knowledge will then help our understanding of potential planetary and cometary responses to transient energy fluxes. Figure 1 shows a view of our "wind tunnel" on 7 March 1970 just prior to some of the largest solar flares during solar cycle 20.

The purpose of this paper is twofold: (1) to summarize some expected solar wind perturbations which may be caused by magnetospheres or ionospheres which are hypothesized to exist around the outer planets; and (2) to discuss some solar wind disturbances (primarily shock waves) which can propagate to the outer solar system.

Magnetosphere- or Ionosphere-Generated Disturbances in the Quiet Solar Wind

Evidence for the utility of fluid continuum analysis for solar wind interaction with Earth's magnetosphere is well documented (see, for example, Wolfe and Intrilagator[5]). This analysis consists of well-known supersonic blunt-body gasdynamics where, in the present context, the interplanetary magnetic field constrains the plasma, via little-understood collective effects, to behave like a continuum fluid in its expansion from the sun as well as in its interaction with obstacles whose characteristic dimension is larger than the proton gyroradius. It was perfectly natural, then, to expect an Earth-like interaction (i.e., magnetospheric shock) following the discovery of shock waves at both Venus and Mars (Fig. 2 and 3, respectively). As noted in Fig. 2, an upper limit for a possible Venusian dipole magnetic moment (as suggested by the shock observations of Venus 4 and Mariner 5) is about 10^{-3} that for Earth (8.06×10^{25} G-cm^3). An alternative explanation (see, for example, Spreiter et al.[6]) of the shock

Table 1   Estimates of the amount of energy expended in the
most important and small flare phenomena (Pinter[4])

	Minimum Energy (erg)	Maximum Energy (erg)
Hα	10^{26}	1.1×10^{31}
Total line emission	-	8×10^{31}
Loop prominences	-	5×10^{31}
Flare blast waves	-	5×10^{30}
Flare surge	-	10^{30}
Flare spray	-	10^{31}
White light flare	3×10^{30}	9.5×10^{30}
EUV burst	$\sim 10^{27}$	$\geq 1.7 \times 10^{31}$
Lyman - burst	-	3.2×10^{30}
Soft X-ray burst (8-20 Å)	10^{29}	5×10^{31}
Soft X-ray burst (8-12 Å)	2×10^{28}	1×10^{31}
Soft X-ray burst (2-12 Å)	2×10^{27}	10^{30}
Soft X-ray burst (1-8 Å)	2×10^{28}	2×10^{29}
Hard X-ray burst (24-4 keV)	1×10^{26}	1×10^{27}
Hard X-ray burst (50-10 keV)	10^{23}	4×10^{27}
Very Hard X-ray burst (90-30 keV)	-	5×10^{25}
Radio burst (3-10 cm)	10^{22}	$\sim 10^{24}$
Type II radio burst	10^{21}	$\sim 10^{25}$
Type III radio burst	10^{20}	10^{21}
Type IV radio burst	10^{22}	10^{25}
Solar electron event > 40 keV	3×10^{25}	4×10^{26}
Solar electron event > 70 keV	4×10^{24}	9×10^{26}
Energetic protons (E > 10 MeV)	-	2×10^{31}
Cosmic ray (1-30 MeV)	-	3×10^{31}
Interplanetary Shock Wave	5×10^{30}	2×10^{32}

Fig. 1   A montage of solar active regions (in Hα, 6563 Å)
and the corona observed before and during (respectively) the total eclipse of 7 March 1970.
The dark circle of the moon has been covered by
the hydrogen-alpha photograph taken in Boulder,
Colorado, by the NASA/NOAA solar patrol telescope.
The corona was photographed during the eclipse by
the High Altitude Observatory, NCAR, which is
sponsored by the National Science Foundation.
(P. S. McIntosh and G. Newkirk, Jr., private communication.)

waves of both Venus and Mars is that they are bow shocks which
are due exclusively to interaction with the ionospheres,
rather than magnetospheres, of these planets. Figure 3 shows
both magnetosphere- as well as the ionosphere-generated shock
waves as proposed, respectively, by Dryer and Heckman[7] and
Spreiter and Rizzi[8] for Mars. It is essential to note that
the mere detection of a planetary bow shock is not a sufficient condition for the existence of either kind of shock
wave. The gasdynamicist will immediately recognize that the
shock wave will be present in a supersonic flow provided the
boundary condition in either case consists in a pressure balance and a turning of the flow such that it becomes tangent to
a given surface, be it a magnetopause or an ionopause.

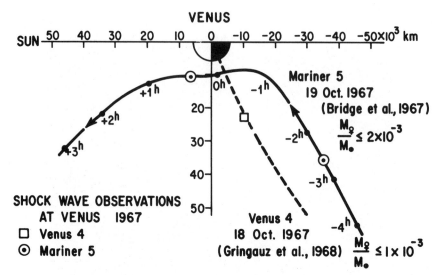

Fig. 2 Observations of a shock wave at Venus
by Soviet and U. S. spacecraft in 1967.
M is the magnetic dipole moment.

The pressure balance, in the magnetospheric case, provides the radial distance, R_p, from the center of the planet
in question to the magnetopause as follows:

$$\frac{R_p}{R_E} = \frac{(M_p/M_E)^{1/3}}{(nV^2)_p/(nV^2)_E}$$ (1)

where R_E is the corresponding, geometrically-scaled distance
at Earth; M_p and M_E are the magnetic dipole moments, again assumed to be scaled accordingly, for the planet and Earth, respectively; and n and V are the average solar wind density and
velocity, respectively, at the planet's orbit and at Earth.

In the case of the purely-ionospheric interaction, the
solar wind pressure is taken to be the Newtonian pressure,
$P = P_{st} \cos^2 \psi$, where P_{st} is the stagnation pressure and ψ is
the angle between the solar wind direction and the normal
vector at the ionopause. The pressure within the ionosphere
is then approximated by the condition of static equilibrium
and set equal to P as follows:

$$P_{st} \cos^2 \psi = P_0 \exp \left[-(r-r_0)/H\right]$$ (2)

where P_o is taken as the pressure at the radial distance from the planet's center to the assumed nose of the ionopause, r_o, along the sun-planet axis. Then, $\cos\psi$ can be expressed in terms of the polar coordinates, r and θ, of the ionopause. H is the scale height of the upper ionosphere, kT/mg, where k is Boltzmann's constant; g is the planet's acceleration of gravity; and T the characteristic temperature of the dominant ionospheric specie of molecular weight, m.

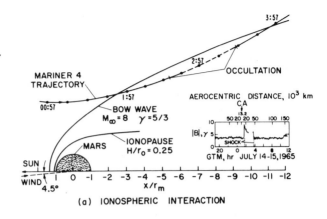

(a) IONOSPHERIC INTERACTION

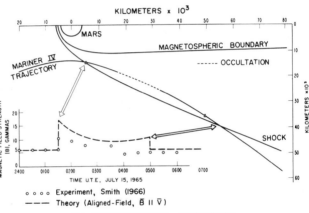

○ ○ ○ ○ Experiment, Smith (1966)
——— Theory (Aligned-Field, $\vec{B} \parallel \vec{V}$)

(b) MAGNETOSPHERIC INTERACTION

Fig. 3 Observations of a shock wave at Mars observed by Mariner 4 in 1965 and alternative suggestions for its existence as being due to either ionospheric (above) or magnetospheric (below) interactions. The latter appears to be correct (see text).

It is presently believed that the ionospheric interaction prevails at Venus, although this point will be examined more closely by the Mariner-Venus-Mercury (MVM) and Pioneer-Venus spacecraft. At Mars, more extensive studies by Mars 2[9] and Mars 3 spacecraft (Dolginov et al.[9] and Gringauz et al.[10] indicate the interaction to be primarily of a magnetospheric character with $M_{Mars}/M_E = 3\times10^{-4}$. Dolginov et al.[9] suggest that such a small dipole moment may be either an ancient field which is a trace of a magnetic dynamo which existed in the past or the signature of a polarity reversal which Mars is

undergoing in its cosmic evolution. Surface measurements,
such as those planned by the U. S. Viking and additional Soviet Mars probes, will help to settle this question.

The interaction at Mercury is presently believed to be
Moon-like because of its small radius (0.38 R_E), slow rotation
(59 days), and apparent lack of an atmosphere. Thus, the solar wind may be absorbed, for the most part, on the sunlit hemisphere and probably produces some limb compressions due to
either boundary layer build-up or even local magnetic anomalies as in Moon's case. The wake probably closes rapidly
after further limb expansion of the high density solar plasma
at Mercury's low heliocentric orbital distance (0.39 astronomical units).

Turning to the first of the non-terrestrial outer planets, Jupiter at 5.2 AU, we are confronted with the in situ
confirmation (December 1973) by Pioneer 10 of a dipolar-like
magnetic field whose magnitude is about $10^4 M_E$. The magnetopause boundary condition is more complex than that given in
Eq. 1 because large fluxes of energetic particles (protons
and electrons) were observed as far as 96 R_J (R_J = Jovian
radius = $11.2 R_E$). This magnetospheric plasma pressure, significant in Jupiter's case, was neglected in the derivation
of Eq. 1 because it is ignorable in Earth's case. The centrifugal force of the rapidly rotating plasma, due to Jupiter's 10 hr rotational period, was similarly neglected. It
is, however, of great significance that a bow shock wave and
magnetopause were observed on both the inbound and outbound
portions of the Pioneer 10 flyby. Preliminary analysis of
quick-look data (Wolfe et al.[11]; Smith et al.[12]) shows that
the solar wind velocity, proton density, proton and electron
temperatures, and magnetic field magnitudes were, respectively: 420 km/sec, $0.03/cm^3$, 7×10^3 °K, 5×10^4 °K, and 0.5Y
(where 1Y = 10^{-5} G). The clearly-identified shock was first
detected at 2030 UT (Earth time) on 26 November 1973 at 109
R_J, about 35° from the Sun-Jupiter axis. The bulk velocity
was reduced immediately downstream in the magnetosheath to
about 250 km/sec; the proton density was compressed to 0.10/
cm^3; the proton temperature increased to 10^6 °K and the magnetic field, to 1.5Y. The magnetopause was detected at 96 R_J
at 2015 UT on 27 November 1973. The magnetopause (detected
again on 1 December) moved inwards, presumably due to a five-
fold increase of solar wind momentum flux detected 7 days
earlier by Pioneer 11 which was located at ~ 3 AU along the
same heliocentric radius. Thus, at 0300 UT on 1 December
1973, the magnetopause moved inside of Pioneer 10 (at 52.5
R_J) and, at 1400 UT, it moved outward again when the space-

craft was at 45.9 R_J (H. Collard, private communication).
During the outbound trajectory, the magnetopause was first
detected at 98 R_J at 1230 UT on 10 December, and the shock, at
124 R_J at ~ 1545 UT on 12 December. The shock and magnetopause were then observed several times again, indicating that
the Jovian magnetosphere was in a state of dynamic motion—a
situation reminiscent of similar observations at Earth during
magnetic storms.

The observation of the Jovian shock wave and its magnetopause confirms the utility, once again, of the supersonic
flow analysis of a continuum fluid. Earlier estimates of the
interaction, based on Eq. 1 for the boundary condition, were
made by Dryer et al.[13] who used the following solar wind parameters (From Cuperman et al.[14]): $n = 0.35/cm^3$, V = 304 km/sec,
$T_p = 8x10^3$ °K, $T_e = 3.2x10^4$ °K, $|B_\infty| = 0.72\gamma$, hence: M=12.9,
$M_A = 11.4$, where M and M_A are the ordinary gasdynamic and
Alfven Mach numbers, respectively, and are not to be confused
with the symbol for the dipole moment. In addition, they assumed $M_J/M_E = 5x10^4$ which, as we now believe after the Pioneer
10 preliminary analysis following the flyby, is a factor of 3
too high. Part of their result is given in Fig. 4 which shows
the estimated magnetopause, bow shock wave, and contours of
normalized isogauss contours within the magnetosheath. Also
shown is the trajectory of a hypothetical Grand Tour space
probe, JSP77 (Jupiter-Saturn-Pluto, 1977 launch), which is
similar to the Pioneer 10 trajectory. The actual trajectory,
given by Greenstadt[15] shows that the spacecraft penetrated
the shock (as noted above) at r = 109 R_J, $\theta \approx 35°$; and first
detected the shock on the outbound trajectory at r = 124 R_J,
$\theta \approx 100°$ (where θ is the sun-planet-spacecraft angle). The
observed magnetic field increase at the shock was 3, which
agrees fairly well with the predicted value as shown in Fig.
4. Agreement of the predicted velocity and density ratios
was also satisfactory. The temperature ratio prediction,
however, was underestimated possibly because of the sensitivity of the latter to the actual Mach number ($T \propto M^2$) which
was more than 20, compared to the assumed value of about 13;
the preliminary observation gives a proton temperature ratio,
$T/T_\infty = 10^6/7x10^3 \cong 140$, whereas the predicted average temperature ratio at the shock entry point was ~ 35. Taken in its
entirety, however, the fluid analysis was, as for Earth and
Mars, remarkably successful. We are, therefore, encouraged
to use this approach for the other outer planets.

At this point, however, we are faced with our ignorance
about the existence of planetary magnetic fields and/or ionospheres at Saturn (9.54 AU), Uranus (19.2 AU), Neptune

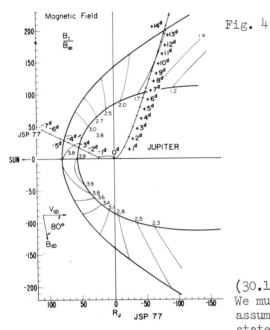

Fig. 4 Predicted bow shock and normalized magnetic field in the magneto-sheath at Jupiter under steady-state conditions. $M_\infty = 10$, where M_∞ is the ordinary Mach number, not to be confused with the magnetic dipole moment. The numbered lightly-drawn lines are values of constant B_1/B_∞, where B_1 is the local field and B_∞ is the ambient magnetic field.

(30.1 AU), and Pluto (39.4 AU). We must, therefore, make some assumptions which, clearly stated, will form the basis of our speculation regarding the type of solar wind interaction which these planets are experiencing. Inasmuch as the boundary condition expressed in Eq. 1 has been successful in its description and approximation of the actual physical situation for Earth, Mars, and now Jupiter, we might make estimates for the possible existence of dipole fields at Saturn, Uranus, and Neptune. These three planets consist mainly of hydrogen and helium. Their atmospheres contain some methane and, possibly, ammonia. Unlike the observations at Jupiter, no radiation belts have been detected by decametric and decimetric radio observations. Thus, the only physical hint regarding the possibility of intrinsic magnetic fields is, like Jupiter, their low rotational periods: 10 hr 24 min for Saturn, 10 hr 50 min for Uranus, and 15 hr 40 min for Neptune. Pluto's rotation period is about 6 days; therefore, like Mercury, it probably has no field.

As to the magnitudes of the dipole moments to be used in our speculative exercise, we turn to a hypothesis discussed by Blackett[16], Moroz[17], and Warwick[18]. This hypothesis states that the angular momentum of a rotating cosmic body is directly proportional to a magnetic dipole moment which is generated, presumably, in a dynamo-like fashion by a highly conducting core. The constants of proportionality (where L_E

and M_E are the Earth's angular momentum and dipole magnetic moment, respectively) for several celestial objects, as an example, are

For Venus: $(L/M)_V = (4.1 \times 10^{-3} L_E)/(\leq 2 \times 10^{-3} M_E)$

$$\geq 1.4 \times 10^{15} \text{ erg sec/G-cm}^3 \tag{3}$$

For Earth: $(L/M)_E = (6 \times 10^{40} \text{ gm-cm}^2/\text{sec})/(8 \times 10^{25} \text{ G-cm}^3)$

$$= 0.7 \times 10^{15} \text{ erg sec/G-cm}^3 \tag{4}$$

For Moon: $(L/M)_{Mn} = (2.5 \times 10^{-5} L_E)/(0.5 \times 10^{-6} M_E)$

$$= 3.5 \times 10^{16} \text{ erg sec/G-cm}^3 \tag{5}$$

For Mars: $(L/M)_M = (2.5 \times 10^{-2} L_E)/(3 \times 10^{-4} M_E)$

$$= 5.8 \times 10^{16} \text{ erg sec/G-cm}^3 \tag{6}$$

For Jupiter: $(L/M)_J = (6.7 \times 10^4 L_E)/(10^4 M_E)$

$$= 0.2 \times 10^{15} \text{ erg sec/G-cm}^3 \tag{7}$$

For the Sun (assuming an average surface field of 1 Gauss):

$$(L/M)_{Sun} = 1.7 \times 10^{48}/0.35 \times 10^{33}$$

$$= 5 \times 10^{15} \text{ erg sec/G-cm}^3 \tag{8}$$

It is seen that the "constant" of proportionality has a range of several orders of magnitude for those celestial objects for which we have a reasonable amount of information. Thus this hypothesis, or "magnetic Bode's law", must be used with great caution. We choose to do so only because the choice of a "constant" makes it possible to estimate an assumed dipole moment for each planet in a consistent fashion. Using this rationale, then, Dryer et al.[13] chose 10^{15} in cgs units, following Warwick[18]. Part of the result for Jupiter was discussed above in connection with Fig. 4. The assumed dipole moment, then, for Saturn is $10^4 M_E$; for Uranus, $2.4 \times 10^2 M_E$; and for Neptune, $1.7 \times 10^2 M_E$. It is further assumed that: (a) the rings of Saturn are, like Moon, essentially non-conducting and will have no effect on the interaction; and (b) the orientation of Uranus' rotational axis of $7.9°$ to its orbital plane will not affect the shape of its magnetosphere as scaled from that of Earth by Eq. 1.

Part of the results found by Dryer et al.[13], then, is given in Fig. 5 for Saturn's normalized temperature ratios in its magnetosheath; in Fig. 6 for Uranus' normalized density ratios; and in Fig. 7 for Neptune's normalized magnetosheath velocity ratios. The hypothetical spacecraft trajectory shown in each case is that for one of the now-cancelled Grand Tour space probes: Jupiter-Saturn-Uranus-Neptune, to be launched in 1977 (i.e., JSUN 77). The mathematical details for the blunt-body MHD computations are given by Shen[19] who considered the oblique magnetic field explicitly in the computations with the region of analysis restricted to the plane containing the magnetic and velocity vectors. It is believed that this latter restriction will still provide reasonably good estimates for regions which differ appreciably from this plane.

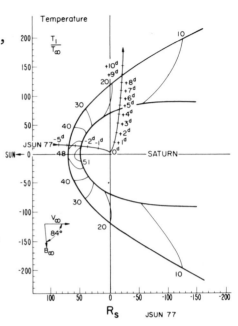

Fig. 5   Predicted bow shock and normalized average temperatures in the magnetosheath of Saturn. $M_\infty = 12.6$.

Let us suppose, alternatively, that Saturn has no magnetic field at all but, instead, has an ionospheric interaction like that of Venus. This possibility, also considered by Dryer et al.[13], is illustrated (with the use of Eq. 2) in Fig. 8 which shows contours of constant density with a hypothesized ionosheath of Saturn. It is important to note that projected trajectories ought to be directed in close proximity of the rings (unlike the JSP 77 trajectory shown in Fig. 8) in order to make definitive identification of a possible ionospheric-generated shock wave. This calculation assumes that the interplanetary magnetic field is parallel to the solar wind, as in an earlier study for Venus and Mars by Spreiter et al.[6]

Finally, Pluto—whose average distance from the Sun in its highly eccentric orbit is 39.4 AU—almost certainly has no magnetic field because of its small radius (0.52 R_E) and

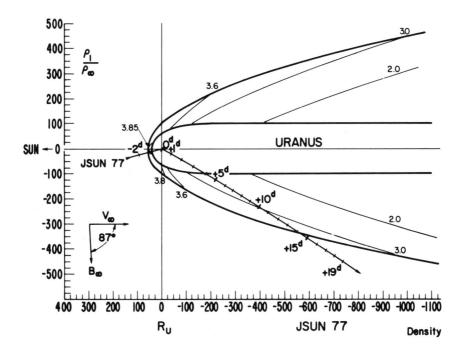

Fig. 6   Predicted bow shock and normalized plasma
densities in the magnetosheath of Uranus.
$M_\infty = 19.1$.

long rotation period of ~ 6 days. Also, no atmosphere has
been detected. Thus we can stretch the continuum analysis
one more step to consider the possible case of a limiting
ionosphere whose scale height, H, approaches zero. As in the
case for Saturn, then, Fig. 9 shows the bow shock for Pluto
as generated by, essentially, the planet itself near the subsolar point and by a small ionopause near the limbs. Contours of both constant velocity and temperature are shown.
Because of the very low temperatures, hence high Mach number,
the ambient average temperature of ~ 2500 $^\circ$K would be increased to nearly coronal values of about 2×10^6 $^\circ$K near the
subsolar region. Still, an alternative interaction possibility is that of near-perfect absorption of the plasma on the
sunlit hemisphere as in the case of the Moon and, possibly,
Mercury. A third possibility—that of a purely kinetic interaction—exists and also deserves consideration (see, for
example, Fig. 10, as suggested by Wu and Dryer[20] for celestial bodies whose size is comparable to or smaller than the
solar wind's characteristic length such as the proton gyroradius).

Solar-Generated Disturbances
in the Solar Wind

The discussion above is, in general, concerned with the
sun as the more-or-less, well-behaved source of a celestial
"wind tunnel" whose steady-state flow is disturbed by a variety of obstacles which happen to have been placed by cosmic evolution in the path of the supersonic plasma. We recognize
that this source is, in reality, inhomogeneous. The sun's active regions (see Fig. 1, for example) produce hot, slow solar
wind outflow. Simultaneously, the darker regions (called
"coronal holes") produce cooler, less dense (but faster) outflow. Because of the sun's rotation, the faster solar wind
will eventually compress the slower plasma which precedes it,
thereby producing a stream-stream interaction. This physical
phenomenon produces the variations of energy and momentum flux
which, as noted above during the discussion of Pioneer 10's
flyby of Jupiter, changes the external boundary conditions for
magnetosphere- and ionosphere-generated disturbances in the
solar wind. The logical consequence of this interaction is an
expansion due to the slow
stream which must follow the
fast stream. Thus, the
planetary environments will
respond by "breathing in
and out" to this external
stimuli. Superimposed
upon this "macroscopic" background is a variety of
Alfvén waves, tangential
and rotational discontin-
uities, and shock waves.
Sometimes the latter are
produced by stream-stream
interactions. The stronger,
more effective shock waves
(from a planetary view-
point) are produced by
solar flares; they are the
subject for the remainder
of this paper. Additional
details are given by Hundhausen[21] and Dryer[22] in
several recent reviews.

Interplanetary shock
waves which are generated by
solar flares are observed first

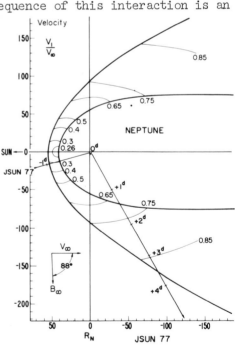

Fig. 7   Predicted bow shock
and normalized bulk
plasma velocities in
the magnetosheath
of Neptune. $M_\infty = 21.0$.

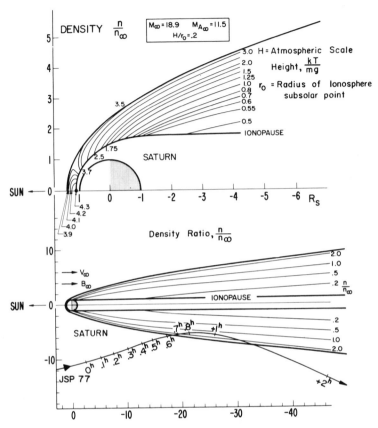

Fig. 8 Alternative ionosphere-generated bow shock wave and
constant density contours with an ionosheath at Saturn.

by ground-based deci- and decametric radio telescopes. These
instruments often detect solar radio emission shortly after
the visible and X-ray portion of some flares. The emission is
characterized by a slow drift from high (~ 250 MHz) to low
(~ 20 MHz) frequencies. More recently, space-borne radio-
meters have extended this diagnostic to lower frequency, decametric and even kilometric wavelengths. This slow drift (referred to as type II) radio emission has a physical explanation based on the "plasma hypothesis". This hypothesis states
that coherent radio emission will occur at a frequency corresponding to the electron plasma frequency, ω, during plasma
oscillations caused by an external forcing function—the shock
wave in this case. Since $\omega = (4\pi e^2 n/m)^{\frac{1}{2}}$; where e, m, and n
are the electron's charge, mass, and density, respectively;
it follows that the motion of the shock wave could be "tracked"
as a function of distance from the sun—provided a coronal
model of electron density is available and folded into the

observations. Average shock velocities of 1500 km/sec have been inferred by this technique. A good example of a "swept-frequency" type II radio burst (from Dulk[23]) is shown in Fig. 11 for a flare on 9 October 1969. A typical type III fast drift (due to coherent beam-plasma instabilities associated with accelerated, relativistic electrons) is seen at 0431.5 UT (flare onset), followed about five minutes later by a split-band, type II fundamental and second harmonic—the latter possibly due to the higher density immediately behind the shock wave. Note that the shock wave was "tracked" for ~ 15 min.

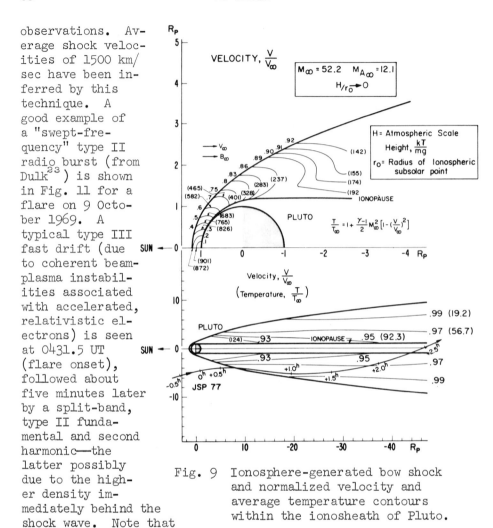

Fig. 9   Ionosphere-generated bow shock and normalized velocity and average temperature contours within the ionosheath of Pluto.

Two-dimensional "photographs" of shock waves have been made possible at discrete frequencies by the radioheliograph in Culgoora, Australia. Figure 12 (from Smerd[24]) shows an 80 MHz radioheliogram (the 2 arc-min resolution and the photosphere are indicated by the dots and an artificial circle, respectively, on the oscilliscope photo) of a series of type II radio bursts. The gross effect, then, is of a shock wave (formed by a series of shocks) frozen by the photograph at 0.6 solar radii above the photosphere. The great extent of the spherical angle of the shock, due to a flare at N19° W110°

just beyond the right limb, is
clearly seen. This instrument's
capability has been extended to
43.25 and 160 MHz, and it is
hoped that the statistics on
events such as this one may be
extended beyond this unique
observation.

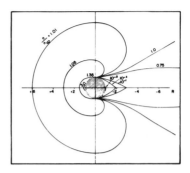

If the energy released by
a flare into the solar wind
takes place during a short interval of time (relative, say,
to the time of 20 to 70 hr
taken for the shock to propagate to Earth's orbit), then
a single shock (or "blast" wave,
as in a nuclear explosion) will be
produced. Should the flare process
(which is still not understood
completely) take place over an extended time, a second compression
process is necessary to adjust to
the previously-produced high

Fig. 10 Plasma kinetic
interaction, as in
a rarefied gasdy-
namic flow, which
is an alternative
possibility for
solar wind inter-
action with Pluto.
Speed ratio$_1$=
$V_\infty/(2kT_\infty/m)^{\frac{1}{2}} = 10.$

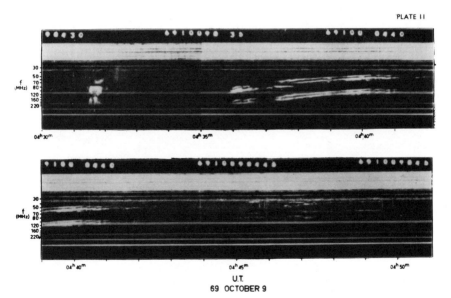

Fig. 11 Swept-frequency spectrum of a type II solar
radio burst following a flare on 9 October
1969.

pressure behind the first (or
"forward") shock wave. This
process is accomplished by a
second or "reverse" shock
wave which is shown in Fig.
13 in schematic form (Greenstadt et al.[25]). The ad-
justment from the original
steady state to either a new
steady or constantly changing state may be stated in a
slightly different way. The
reverse shock is convected
away from the sun although
an observer in the shock's
frame of reference will observe the plasma coming
from the sun at a velocity
greater than his own. This
second shock wave produces
a pressure increase which
is required to match that
caused at the piston (dashed line in the sketch) by
the first (forward) shock
wave. The gross effect of
this second kind of flare,
then, is to cause a signi-

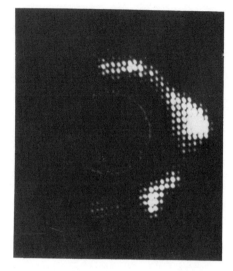

Fig. 12   Radioheliogram (80
          MHz) of solar flare-
          generated shock waves
          obtained at 0250 UT
          on 30 March 1969 fol-
          lowing a flare just
          beyond the west
          (right) limb.

ficant modulation of the interplanetary magnetic field,
$\overline{B}$, and a plasma compression between the two asymmetrical out-
ward-propagating shock waves. The nature of the shock structure itself may change from a "perpendicular" to "parallel"
shock as the observing plasma detector or magnetometer moves
from a position west of the flare's central meridian (CM) to
another position east of the flare's CM. The definition of
these "collisionless" shock structures refers to the angle
between the upstream magnetic field vector and the normal to
the shock surface. Thus, as suggested in the lower part of
Fig. 13, the monotonic magnetic field increase through the
shock indicates that the latter's thickness (about 10^3 km) may
be of the order of ten ion inertial lengths, where c is the
velocity of light and ω_{pi} is the ion plasma frequency. The
multi-gradient shock, on the other hand, is poorly-defined and
may, therefore, be less effective in its perturbation of, say,
a magnetosphere, ionosphere, or comet. The dashed line between the two shocks represents a contact discontinuity, or
"piston", which marks the existence of enhanced plasma and
magnetic compression.

The piston is illus-
trated in another way by
using either similarity
theory or numerical simu-
lations for multiple shock
propagation studies. A
representative result
from similarity theory
(Dryer et al.[26]; Dryer[27])
is shown in Fig. 14. Note
that—as suggested in
Fig. 13—a strong modula-
tion of the interplanetary
field takes place with a
concomitant scattering
effect on both incoming
galactic and outgoing
solar cosmic rays. The
lower half of the figure
shows a cross-section of
the density pulse (at 14
and 52 hours after the
flare) which connects the
original (dashed) solar
wind density to the final
(dotted) state. The dis-
continuity (roughly half-
way between the forward
and reverse shocks) at the

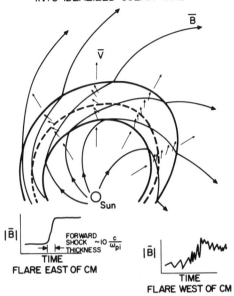

Fig. 13 Schematic propagation
of a multiple shock
ensemble through the
solar wind.

piston reflects the fact that (at least for the assumption of
infinite electrical conductivity in this example) momentum
transfer at this location takes place purely by magnetic ten-
sion and not by thermal pressure. In reality, steep gradients
in density (and field) as shown in Fig. 14 would not be likely
to be sustained due to magnetic drift wave instabilities (Unti
et al.[28]).

Obviously, multiple observation points (in the ecliptic)
for a given event are clearly required in order to confirm or
refute any of the physical assumptions extant in any of the
analytical or numerical time-dependent solutions. Out-of-
ecliptic observations, such as those discussed by Wilcox[29],
are also necessary in order to determine shock dynamics in the
third dimension. Realistically, however, we must be content
to exploit ecliptic observations as much as possible.

Such an opportunity came during a clearly definitive
series of large solar flares during August 1972—the most

important activity
during solar cycle
20. The observation
points, relative
to a fixed sun-
Earth axis, are
shown (in the eclip-
tic plane) in Fig.
15. They include
Earth (in the form
of sudden commence-
ments of magnetic
storms), Pioneers 9
and 10, and the
periodic Comet
Giacobini-Zinner.
The latter was very
close to or in the
ecliptic plane
about the time of
its perihelion on
3 August 1972.
Significant bright-
ness modulations
following perihelion
are believed (Dryer
et al.[30]) to be
caused by impact of
the greatly-enhanced,
flare-generated solar

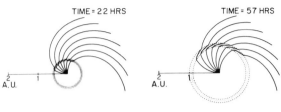

Magnetic Topology

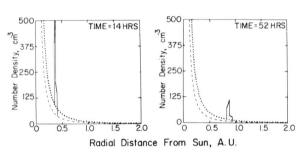

Radial Distance From Sun, A.U.

Fig. 14   Theoretical double-shock
ensemble as viewed in the
ecliptic plane (magnetic
topology deformation) and
along a radial vector from
the sun (density pulse).

plasma with the resonantly-fluorescing cometary radicals. On
a short time scale of, say, 10 hr the net effect could be a
temporary decrease in visual brightness. As a result, the
comet could be used as a natural probe of the solar wind.
Jupiter and Comet Schwassmann-Wachmann I, indicated in Fig.
15, did not respond to the solar activity, probably because
the shock waves (from the early flares) decelerated to Alfvén
waves prior to 5 AU and were too weak along their flanks for
the shock from the 7 August 1972 flare (as indicated by the
dashed line in Fig. 15).

The series of flares began on 2 August 1972 in McMath
Region 11976 at $N14^\circ$ $E35^\circ$, continued on 4 August at $N15^\circ$ $E09^\circ$,
and on 7 August at $N14^\circ$ $W38^\circ$. The shock trajectory from this
last flare is of particular interest because it was tracked
from the Sun to most of the points indicated in Fig. 15. The
type II radio drift was measured[31] down to 30 kHz (hence, as
far as Earth) by a radiometer on Imp 6. Figure 16 shows the

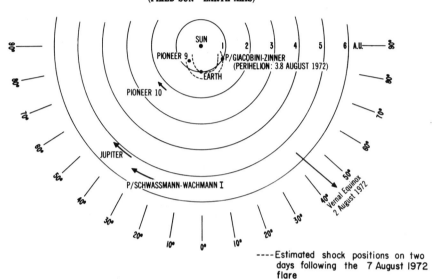

Fig. 15   "Observation points" in the ecliptic plane
          during the solar activity of 2-11 August 1972,
          relative to a fixed Sun-Earth axis.

average velocity during its transit of 1 AU to be 1270 km/sec,
which corresponds to the average velocity (point labeled S-E)
between the optical flare and the sudden commencement of a
major geomagnetic storm. The density model used for this observation is based on the second harmonic type III measurements by the RAE-1 satellite. Average velocities between
other stations such as Sun-Pioneer 9 (S-9), Pioneer 9-Pioneer
10 (i.e., 9-10), etc., are estimated under the assumption of
spherical propagation. The shock observations at Pioneers 9
and 10 were made by Mihalov et al.[32] Hence, the estimate
shown in the figure for the power law index for the shock velocity beyond ~ 0.5 AU must be considered to be an approximation only. It is clear, however, that the shock had a piston-
driven character nearly to 1 AU and a blast-like deceleration
thereafter. As suggested by the estimated straight-line fit
in Fig. 16, the shock degraded into an MHD or Alfvén wave
prior to reaching the vicinity of Jupiter and P/Schwassmann-
Wachmann I. These events are currently being studied in
greater detail.

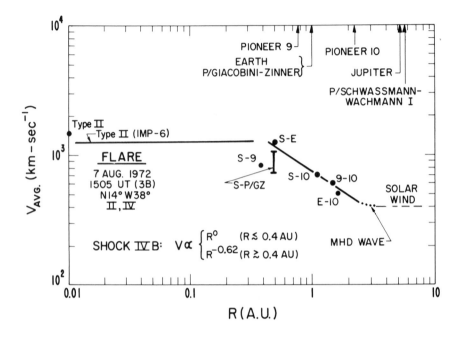

Fig. 16   Average interplanetary shock velocities from
          the flare on Aug. 7, 1972 from McMath Region
          11976. The points, 9-10, S-E, etc., refer to
          the average velocity between Pioneers 9 and
          10, Sun and Earth, etc.

Conclusions

     Continuum fluid mechanics has been utilized to study:
(a) the steady-state supersonic interaction of the solar wind
with planetary magnetospheres and ionospheres, and (b) the
time-dependent propagation of solar flare-generated shock
waves through the solar wind. The former study has been successful for Earth, Mars, and possibly Venus and Jupiter.
Speculations are thereby extrapolated to the other outer
planets in the solar system. The time-dependent studies of
interplanetary shock waves look very promising but require
additional comparisons of observations and theory.

     Note added in proof: Several developments have taken
place since this paper was originally prepared. A bow shock
and magnetosphere have been discovered at Mercury[33]. Also,
interplanetary studies have shown important progress[34,35].

References

[1] Evans, J. W., "Introductory Review of Solar Activity," AIAA Progress in Astronautics and Aeronautics: Solar Activity Observations and Predictions, ol. 30, edited by P. S. McIntosh and M. Dryer, MIT Press, Cambridge, Mass., 1972, pp. 3-17.

[2] Allen, C. W., "The Interpretation of the XUV Solar Spectrum," Space Science Reviews, Vol. 4, No. 1, 1965, pp. 91-122.

[3] Reid, G. C., "Ionospheric Effects of Solar Activity," AIAA Progress in Astronautics and Aeronautics: Solar Activity Observations and Predictions, Vol. 30, edited by P. S. McIntosh and M. Dryer, MIT Press, Cambridge, Mass., 1972, pp. 293-312.

[4] Pintér, S., "Energy Content of Solar Flares," in Proceedings of Leningrad International Seminar, Nauka, Moscow, 1972, pp. 64-100.

[5] Wolfe, J. H., and Intriligator, D. S., "The Solar Wind Interaction with the Geomagnetic Field," Space Science Reviews, Vol. 10, No. 4, 1970, pp. 511-596.

[6] Spreiter, J. R., Summers, A. L., and Rizzi, A. W., "Solar Wind Flow Past Non-Magnetic Planets-Venus and Mars," Planetary and Space Science, Vol. 18, 1970, pp. 1281-1299.

[7] Dryer, M., and Heckman, G. R., "Application of the Hypersonic Analogue to the Standing Shock of Mars," Solar Physics, Vol. 2, No. 1, 1967, pp. 112-124.

[8] Spreiter, J. R., and Rizzi, A. W., "The Martian Bow Wave-Theory and Observation," Planetary and Space Science, Vol. 20, No. 2, 1972, pp. 205-208.

[9] Dolginov, Sh. Sh., Yeroshenko, Ye. G., and Zhuzgov, L. N., "Magnetic Field in the Very Close Neighborhood of Mars According to Data from the Mars 2 and Mars 3 Spacecraft," Journal of Geophysical Research, Vol. 78, No. 22, 1973, pp. 4779-4786. Also: Doklady Akademiia Nauk SSR, Vol. 207, 1972, pp. 1296-1299.

[10] Gringauz, K. I., Bezrukikh, V. V., Volkov, G. I., Breus, T. K., Musatov, I. S., Havkin, L. P., and Sloutchenkov, G. F., "Results of Solar Plasma Electron Observations

on Mars-2 and Mars-3 Spacecraft," Journal of Geophysical Research, Vol. 78, No. 25, 1973, pp. 5808-5812.

[11] Wolfe, J. H., Collard, H. R., Mihalov, J. D., and Intriligator, D. S., "Preliminary Pioneer 10 Encounter Results from the Ames Research Center Plasma Analyzer Experiment," Science, Vol. 183, No. 4122, 1974, pp. 303-305.

[12] Smith, E. J., Davis, Jr., L., Jones, D. E., Colburn, D. S., Coleman, Jr., P. J., Dyal, P., and Sonett, C. P., "Magnetic Field of Jupiter and its Interaction with the Solar Wind," Science, Vol. 183, No. 4122, 1974, pp. 305-306.

[13] Dryer, M., Rizzi, A. W., and Shen, W. -W., "Interaction of the Solar Wind with the Outer Planets," Astrophysics and Space Science, Vol. 22, No. 2, pp. 329-351.

[14] Cuperman, S., Harten, A., and Dryer, M., "Characteristics of the Quiet Solar Wind Beyond the Earth's Orbit," Astrophysical Journal, Vol. 177, Part 1, 1972, pp. 555-566.

[15] Greenstadt, E. W., "Oblique Structure of Jupiter's Bow Shock," Journal of Geophysical Research, Vol. 78, No. 25, 1973, pp. 5813-5817.

[16] Blackett, P. M. S., "The Magnetic Field of Massive Rotating Bodies," Nature, Vol. 159, No. 4046, 1947, pp. 658-666.

[17] Moroz, V. I., Physics of Planets, Moscow, 1967; (English Translation: NASA Technical Translation F-515, 1968).

[18] Warwick, J. W., "Particles and Fields Near Jupiter," NASA Report CR-1685, 1970.

[19] Shen, W. -W., "The Earth's Bow Shock in an Oblique Interplanetary Field," Cosmic Electrodynamics, Vol. 2, No. 4, 1972, pp. 381-395.

[20] Wu, S. T., and Dryer, M., "Kinetic Theory Analysis of Solar Wind Interaction with Planetary Objects," in Photon and Particle Interactions with Surfaces in Space, edited by R. J. L. Grard, D. Reidel Publishing Company, Noordwijk, Holland, 1973, pp. 453-470.

[21] Hundhausen, A. J., Coronal Expansion and Solar Wind, Springer-Verlag, New York, 1972.

[22] Dryer, M., "Interplanetary Shock Waves Generated by Solar Flares," _Space Science Reviews_, Vol. 15, No. 4, 1974, pp. 403-468.

[23] Dulk, G. A., "Positions of the Fundamental and Harmonic Sources of a Type II Solar Burst," _Proceedings, of the Astronomical Society of Australia_, Vol. 1, No.7, 1970, pp. 308-310.

[24] Smerd, S. F., "Radio Evidence for the Propagation of Magnetohydrodynamic Waves along Curved Paths in the Solar Corona," _Proceedings of the Astronomical Society of Australia_, Vol. 1, No. 7, 1970, 305-308.

[25] Greenstadt, E. W., Dryer, M., and Smith, Z. K., "Field-Determined Structure of Interplanetary Shocks," in _Proceedings of the Conference on Flare-Produced Shock Waves in the Corona and Interplanetary Space_, edited by A. J. Hundhausen, National Center for Atmospheric Research, 1974, pp. 245-265.

[26] Dryer, M., Smith, Z. K., Endrud, G. H., and Wolfe, J. H., "Pioneer 7 Observations of the August 29, 1966, Interplanetary shock-Wave Ensemble," _Cosmic Electrodynamics_, Vol. 3, No. 2, 1972, pp. 184-207.

[27] Dryer, M., "Interplanetary Double-Shock Ensembles with Anomalous Electrical Conductivity," in _Solar Wind_, edited by C. P. Sonett, P. J. Coleman, Jr., and J. M. Wilcox, NASA SP-308, Supt. of Documents, Washington, 1972, pp. 453-465.

[28] Unti, T., Neugebauer, M., and Wu, C. -S., "Shock System of February 2, 1969," _Journal of Geophysical Research_, Vol. 78, No. 31, 1973, pp. 7237-7256.

[29] Wilcox, J. M., "Scientific Exploration with an Out-of-Ecliptic Spacecraft," Institute for Plasma Research, Stanford University, Report No. 504, 1973.

[30] Dryer, M., Eviatar, A., Frohlich, A. L., Jacobs, A., Joseph, J. H., and Weber, E. J., "Interplanetary Shock Waves from McMath Region 11976 During its Passage in August 1972," in _Coronal Disturbances, Proceedings of IAU Symposium No. 57_, edited by G. Newkirk, Jr., D. Reidel Publishing Co., Dordrecht, 1974, pp. 377-381.

[31] Malitson, H. H., Feinberg, J., and Stone, R. G., " A Density
     Scale for the Interplanetary Medium from Observations
     of a Type II Solar Radio Burst Out to 1 Astronomical
     Unit," Astrophysical Journal, Vol. 183, Part 2, 1973,
     pp. L35-L38.

[32] Mihalov, J. D., Colburn, D. S., Smith, B. F., Sonett, C. P.,
     and Wolfe, J. H., "Pioneer Solar Plasma and Magnetic
     Field Measurements in Interplanetary Space during Aug-
     ust 2-17, 1972," in Correlated Interplanetary and Mag-
     netospheric Observations, Proceedings of the Seventh
     ESLAB Symposium, edited by D. E. Page, D. Reidel Publ.
     Co., Dordrecht, 1974, pp. 545-553.

[33] Hartle, R. E., Ogilvie, K. W., Scudder, J. D., Bridge, H. S.,
     Siscoe, G. L., Lazarus, A. J., Vasyliunas, V. M., and
     Yeates, C. M., "Preliminary Interpretations of Plasma
     Electron Observations at the Third Encounter of Mariner
     10 with Mercury," Nature, Vol. 255, No. 5505, 1975,
     pp. 206-208.

[34] Dryer, M. "Interplanetary Shock Waves Generated by Solar
     Flares," Space Science Reviews, Vol. 15, No. 4, 1974,
     pp. 403-468.

[35] Dryer, M., "Interplanetary Shock Waves: Recent Developments,"
     Space Science Reviews, Vol. 17, Nos. 2/3/4, 1975, pp.
     277-325.

CHAPTER II—MAJOR PLANETS

GRAVITATIONAL FIELDS AND INTERIOR
STRUCTURE OF THE GIANT PLANETS

John D. Anderson[*]
California Institute of Technology, Pasadena, Calif.

and

William B. Hubbard[+]
University of Arizona, Tucson, Ariz.

Abstract

A review of the analysis and interpretation of gravity
data from the Pioneer 10 flyby of Jupiter in December 1973 is
presented. The relationship between the external gravitational
field of a giant planet and the distribution of matter in its
interior is discussed in terms of a new theory of gravity
sounding. The objective of this review is to provide an elementary understanding of the information contained in gravitational data for purposes of planning future planetary missions
and for purposes of anticipating what will be learned from
future flybys with Pioneer 11 and the Mariner Jupiter/Saturn
spacecraft.

Introduction

Analysis of two-way coherent Doppler data from spacecraft
that flyby or orbit the giant planets will provide in the next
few years definitive measurements on the gravity fields of

Presented at AIAA/AGU Space Science Conference: Exploration
of the Outer Solar System, Denver, Colorado, July 10-12, 1973/
(not preprinted). The work presented in this paper represents
one phase of research at the Jet Propulsion Laboratory, Calif-
ornia Institute of Technology, under NASA Contract NAS 7-100.
Hubbard acknowledges the support of NASA Grant NSG-7045.
*Staff Scientist, Mission Analysis Division, Jet Propulsion
Laboratory.
+Professor, Department of Planetary Sciences, Lunar and
Planetary Laboratory.

71

those planets. Such measurements are of significance in that
they tell planetologists a great deal about the internal structure of the planet. This is so because measurements of the
gravitational field, when combined with knowledge on the physical properties of hydrogen and helium at high pressures and
temperatures, can be used to draw conclusions on the physical
properties of a giant planet. For example, data from the
Pioneer 10 flyby of Jupiter in December 1973 have been used to
derive a structure for the outer envelope of the planet which
is consistent with an adiabatic, solar-composition envelope
with a starting temperature of (250 ± 40)K at a pressure of 1
bar.[1]

The relationship between the interior of a giant planet
and its gravity field is dependent on a rapid rotation rate. If
the planet did not rotate, it would take on a spherical shape
under its own self-gravitation. For purposes of this discussion, tidal effects produced by the sun and other bodies can
be neglected, and, as far as any gravity-sensing experiment is
concerned, a nonrotating planet would appear essentially as a
point mass. The external gravitational field would be spherical for all of the possible radial density distributions, and
it would be impossible to infer anything about the
density distribution from the gravity data. However, because
the planet rotates, its shape will differ from a sphere, and
the amount of the deviation from sphericity will be reflected
in the external gravity field. The amount of the deviations
will depend on the density distribution within the planet. For
example, if the mass of the planet were concentrated completely
at the center, then it would behave as a point mass, and the
measurements of the external gravity field would yield a
spherical structure. On the other hand, if the planet were
homogeneous, the deviations from sphericity would be at a maximum, under the assumption that the density does not decrease
with depth, and this maximum deviation would be evident in the
gravity data. The actual situation for the giant planets falls
somewhere between the two extremes of total concentration at
the center and a homogeneous planet. Accurate measurements of
the gravitational field can contribute significantly to a
specification of exactly how the density varies with depth.

Because the planets can be treated as spheres to a zero-
order approximation, it makes sense to express their external
gravity fields in terms of spherical harmonics. Furthermore,
the giant planets are assumed to be in hydrostatic equilibrium,
and as a result all of the spherical harmonics except the even

zonals (J_2, J_4, J_6, ...) are zero. Therefore, the gravitational potential can be written in the usual form:

$$U = \frac{GM}{r} \left[1 - \sum_{\ell=1}^{\infty} J_{2\ell} \left(\frac{R}{r} \right)^{2\ell} P_{2\ell} (\sin\phi) \right] \qquad (1)$$

where M is the total mass of the planet, R is its equatorial radius, r is the distance from the center of mass of the planet to a point in space, and ϕ is the latitude of the point with respect to the equatorial plane of the planet. The potential function is defined such that its gradient will yield the equations of motion for a test particle.

The current technique for measuring the gravitational field of a giant planet is to observe the motion of a spacecraft by means of accurate two-way coherent Doppler data, and then to determine the best values of the coefficients $J_{2\ell}$ that will reproduce that motion. The close approach of Pioneer 10 to Jupiter at a distance of about 2.8 Jupiter radii, coupled with Doppler measurements accurate to 5 mHz (0.3 mm/sec) over a count time of 60 sec, has yielded the first definitive measurement of J_4 and has determined J_2 with considerably more accuracy than that obtainable from the motions of the Galilean satellites. The results of the Pioneer 10 analysis are[2]

$$J_2 = (1.4720 \pm 0.0040) \times 10^{-2}$$

$$J_4 = (-6.5 \pm 1.5) \times 10^{-4}$$

where the values of the coefficients are based on an assumed radius R of 71,400 km.

The significance of a definitive measurement of the harmonic coefficients $J_{2\ell}$ for Jupiter, or for any other giant planet for that matter, rests on the relationship

$$J_{2\ell} = - \frac{1}{MR^{2\ell}} \int_V \rho(r, \phi, \lambda) r^{2\ell} P_{2\ell}(\sin\phi) \, dV \quad \ell = 1,2,3,.. \qquad (2)$$

where $\rho(r, \phi, \lambda)$ represents the density distribution within the planet and the integration is carried out over the entire volume V. A specification of values for the coefficients $J_{2\ell}$ will impose integral constraints on the allowable density distributions, but it should be noted that it is impossible to determine a unique density distribution from a finite number of gravity coefficients. In this sense, the observed coefficients impose necessary but not sufficient conditions on the validity of any proposed planetary model. A complete model for a giant planet, in addition, depends on a knowledge of the chemical composition of the planet and on an understanding of the equation of state of the planetary material, which may be under pressures in the 10-100 Mb region and be at temperatures of several thousand degrees. Eventually, reliable interior models will depend on accurate determinations of at least J_2, J_4, and J_6 for the giant planets. The coefficient J_2 yields information on the overall density distribution, including the distribution in the deep interior, whereas the coefficients J_4 and J_6 provide detailed information on the distribution of material in the outer envelope of the planet. A combination of information on the deep interior with detailed information on the outer envelope produces a total picture of the conditions within the planet.

Gravity Fields of the Giant Planets

At this time, only Jupiter has been probed by spacecraft, and consequently knowledge on the gravity fields of Saturn, Uranus, and Neptune must be obtained from the motions of their natural satellites. By observing the advance of the pericenter and the regression of the nodes of satellite orbits over long periods of time, it is possible to determine values of J_2 and J_4 which produce the observed motions. A thorough discussion of this technique has been given by Brouwer and Clemence.[3] It works most successfully for Saturn, where the motions of the six inner satellites yield values for both J_2 and J_4. However, it is a difficult technique to apply to the Jupiter system, where the Pioneer 10 flyby can provide far superior results. Recently, Whitaker and Greenburg[4] have remeasured all available plates showing Uranus' fifth satellite, Miranda, which was discovered by Kuiper in 1948, and have concluded that J_2 for Uranus must be in the neighborhood of 0.005. No information is available on J_4 for Uranus. Neptune has only one satellite, Triton, which can yield information on the gravity field of the planet, but because the satellite orbit is nearly circular at a distance of almost 16 planetary radii, only J_2 can be determined.

The current knowledge on the gravity fields of the giant planets is summarized in Table 1. A recent determination of J_2 and J_4 for Saturn by Garcia[5] has not been included because his positive value for J_4 (0.0014) would imply that the density of material in Saturn is decreasing with increasing depth below the surface of the planet. The wide difference between Garcia's value for J_4 and the value in Table 1 can be viewed as indicative of the difficulty in determining the gravity field

Table 1   Second and fourth degree zonal harmonic
coefficients for the giant planets

Planet	$J_2 \times 10^3$	$J_4 \times 10^4$	Source
Jupiter	14.720 ± 0.040	-6.5 ± 1.5	Pioneer 10[2]
Saturn	16.67 ± 0.02	-10.3 ± 0.8	Brouwer & Clemence[3]
Uranus	5	?	Whitaker & Greenberg[4]
Neptune	4.9 ± 0.5	?	Brouwer & Clemence[3]

of a giant planet from Earth-based optical observations of its satellites. The tracking of spacecraft while they are close to the planet offers the best opportunity for an unambiguous determination of the even-zonal harmonics and, in addition, offers the only known means to detect other coefficients such as J_3, C_{22}, and S_{22}, which would measure deviations from hydrostatic equilibrium.

Planetary Interiors

Because the giant planets are large and probably fluid, it is a good assumption that their structure is dominated by gravitational forces and that they are in hydrostatic equilibrium. Furthermore, their rapid rotation rates will produce significant deviations from sphericity. A useful measure of the rotation is provided by the dimensionless parameter q, which represents the ratio of the centrifugal to gravitational force on the equator at the surface of the planet:

$$q = \omega^2 R^3 / GM \qquad (3)$$

The angular velocity of rotation ω is assumed uniform throughout the body of the planet. The value of q is 0.08885 for Jupiter, 0.1723 for Saturn, 0.0735 for Uranus, and 0.027 for Neptune.

To the first order in q, the equation of hydrostatic equilibrium is

$$\frac{dp}{ds} = -\frac{G\rho(s)\,M(s)}{s^2} + \frac{2}{3}\omega^2 s\,\rho(s) \qquad (4)$$

where p is the pressure, ρ the density, and M the mass as a function of depth:

$$M(s) = 4\pi \int_0^s \rho(a)\, a^2\, da \qquad (5)$$

The independent variable s is associated with the equipotential surfaces of constant density within the planet and is related to the radius r and latitude ϕ by

$$r = s\left[1 + \varepsilon_o(s) + \varepsilon_2(s)\, P_2(\sin\phi) + \varepsilon_4(s)\, P_4(\sin\phi) + \ldots\right] \quad (6)$$

The functions $\varepsilon_i(s)$ are determined by the density distribution $\rho(s)$ and are found by numerically solving a system of integro-differential equations. This solution leads to a specification of the shape of the planet at its surface (s = s_1), and hence the gravitational harmonics (J_2, J_4, J_6,...)are determined as well by Eq. (2). A numerical solution to Eq. (2) has been developed to the third order by Zharkov and Trubitsyn[6] which is valid for any density distribution $\rho(s)$.

The density distribution can be found to the first order in q by numerically integrating Eq. (4) and (5), and then J_2 can be calculated to the second order in q and the coefficients J_4 and J_6 to the third order. Agreement between the calculated values of the harmonic coefficients and the observed values must be achieved before the function $\rho(s)$ can be taken seriously. The overwhelming difficulty with all of this is that a relationship between the pressure and density must be added to Eq. (4) and (5) in the form of an equation of state:

$$p = f(\rho, T) \qquad (7)$$

Furthermore, the fact that Jupiter and Saturn radiate more energy into space than they receive from the sun[7] implies that temperatures on the order of several thousand degrees exist in their interiors. Therefore, the equation of state must include thermal effects on the planetary material (T $\neq$ 0). In addition, the introduction of a new variable T requires an independent thermal relation of the general form

$$T = g(p, \rho) \qquad (8)$$

Note that Eq. (4) and (5) are dynamical and mass continuity equations, respectively, and neither of them depends explicitly on the chemical composition of the planet. However, both Eq. (7) and (8) depend on the chemical composition and on the physical conditions of the planetary material. Therefore, a

realistic model of a planetary interior cannot be constructed
unless the chemical composition is known and unless the physics
of the material at high pressures, densities, and temperatures
is understood. For example, the simplest thermal relation is
T = const, but it is unlikely that this is valid except
possibly in a small rocky core at the center of the planet.
Probably it is closer to reality to assume that the giant planets are completely convective and that the temperature gradient is adiabatic throughout their interiors. For a perfect
gas, the adiabatic relationship corresponding to Eq. (8) is[8]

$$T = c \rho^{\gamma-1} \qquad (9)$$

where c is a constant and γ is the ratio of specific heats of
the gas at constant pressure and constant volume. Of course,
the material can be approximated by a perfect gas only in the
outer levels of the atmosphere. However, it can be shown that,
in the deep interior, where hydrogen will be in a liquid-
metallic state, the adiabatic relationship given by Eq. (9)
still is valid if γ is set equal to 1.5. Nearer the surface,
where hydrogen is in a liquid-molecular state (H_2), the calculation of the adiabatic temperature gradient, along with the
equation-of-state of the molecular hydrogen, is quite complicated. The reader is referred to a recent paper by Podolak
and Cameron[9] for the details of this calculation and also for
an account of the equations-of-state in the interior.

The Interior of Jupiter

The pioneering work on interior models of the giant
planets was performed by De Marcus in 1958.[10] He showed that
Jupiter and Saturn must be made up primarily of hydrogen.
Later Peebles[11] derived more detailed models for Jupiter and
Saturn which yielded the observed values of J_2 and J_4 as given
by Brouwer and Clemence.[3] However, prior to the discovery of
the thermal emission for these two planets, it was assumed
that they were cold and that thermal perturbations to the
equations of state of the hydrogen-helium mixture would be
insignificant. About five years ago, models were constructed
by Hubbard[12, 13] which took into account the thermal perturbations on the equation of state. Since then, theoreticians
have concentrated on trying to understand the thermal effects
on the interior, within the context of reasonable chemical compositions. The most recently published models are those of
Podolak and Cameron.[9]

It should be obvious by now that the detailed study of
the structure of the giant planets is a fairly new discipline,

and undoubtedly much will be learned within the next decade on both the theoretical and experimental sides of the problem. At the present time, it probably makes sense to limit serious discussion of interiors to Jupiter, the only planet that has been probed with spacecraft. Uncertainties in the gravitational harmonics for Saturn make models for that planet correspondingly uncertain, and more severe uncertainties for Uranus and Neptune, not only in their gravity fields but also in their rotation rates, size, and mass, make models for those two planets very speculative. New models for Jupiter are in progress, and a successful return of data from Pioneer 11 at its flyby distance of 1.6 Jupiter radii from the center will provide even more definitive data in the near future. However, for now, a combination of theoretical calculations, along with the Pioneer 10 gravity data, yields the following general model for the interior structure of Jupiter.

The chemical composition of Jupiter probably is the same as the sun, with perhaps some enrichment of methane, ammonia, and water because of a condensation process during the early formation of the planet.[9] Under this assumption, about 75% of the mass of the planet is hydrogen, and the ratio of the abundance of hydrogen to helium by mass is about 3.4. The planet is almost certainly liquid throughout its interior and probably is totally convective.

The only exception to this is that it may contain a small rocky core, perhaps enriched with iron, at a central temperature of about 25,000 K or perhaps somewhat less. Outside of this possibility, the liquid body of the planet consists of two main zones. The inner zone is mainly liquid-metallic hydrogen, and it extends to a radius of about 46,000 km from the center. The remaining 25,000 km or so of the planet consists mainly of liquid-molecular hydrogen. At the transition from metallic to molecular hydrogen, the temperature of the material is about 11,000 K, and the pressure is 3 Mb or about 3×10^6 Earth atmospheres. On top of the liquid body of the planet, there is a gaseous atmosphere with a thickness of about 1000 km from its base to the top of the visable clouds. Because of large-scale convection, the chemical composition of the planet probably is homogeneous, except for the possibility of a small rocky core, and the solar hydrogen-to-helium ratio is maintained throughout.

The Outer Envelope of Jupiter

A definitive measurement of J_2 and J_4 from Pioneer 10 has made it possible to determine empirically the density distribution of material in the outer envelope of Jupiter to a depth of about 3100 km. This determination has been accomplished by means of a new gravitational inversion technique developed by Hubbard.[14] It is assumed that the planet is in hydrostatic equilibrium and that the density near the surface varies smoothly with depth. Under these conditions, it is permissable to expand the density in a power series of the form

$$\rho(s) = \rho_o + \rho_o{'} (s - b) + (1/2) \rho_o{''} (s - b)^2 + \ldots \quad (10)$$

where s is the independent parameter defined earlier, and b is the polar radius of the planet. An approximate solution to Eq. (2) has been found for this quadratic density function which, by inversion, permits a calculation of ρ_o, $\rho_o{'}$, and $\rho_o{''}$ from observed values of J_2, J_4, and J_6. The results for ρ_o $\rho_o{'}$ are

$$\rho_o (s = b) \approx - \frac{35}{4\pi} \frac{J_4}{\left[J_2 + (q/3) \right]} \frac{M}{b^3} \quad (11)$$

$$\rho_o{'} (s = b) \approx \frac{35}{4\pi} \frac{J_4}{\left[J_2 + (q/3) \right]^2} \frac{M}{b^4} \quad (12)$$

Thus, with values of J_2 and J_4 from Pioneer 10, it is possible to determine a linear approximation to the actual density distribution in the outer envelope. The second derivative may be estimated from a knowledge of J_6, which is not yet available for Jupiter. However, under the assumption of linearity, which is not a bad assumption for an adiabatic envelope of solar composition, the density and density gradient can be computed on a level surface characterized by s = b. Then the equations of hydrostatic equilibrium [Eq. (4)] and the mass continuity equation [Eq. (5)] can be integrated to yield an empirical pressure-density profile as a function of s. The surface of the planet where $s = s_1$ is defined for this purpose as the level surface where the pressure reaches a value of 1 bar. The quantity $(s_1 - b)$ then becomes the depth that is being probed by the gravity-sensing experiment. We call this depth the gravitational sounding level at about 3100 km for Jupiter and about 3600 km for Saturn, based on the values of J_2 and J_4 given in Table 1. At this depth, the adiabatic equation of state can be approximated by

$$p = K \rho^2 \quad (13)$$

where K is a constant that depends on the starting temperature of the adiabat at 1 bar pressure and also on the chemical com-

position. By differentiating Eq. (13) and by setting the
result equal to Eq. (4) at s = b, an empirical value for K can
be derived:

$$K = -\frac{2\pi G}{35} \frac{\left[(J_2 + (q/3)\right]^2}{J_4} b^2 \tag{14}$$

On the level surface defined by s = b, the actual equation of
state should osculate the empirical relation given by Eq. (13)
and (14).

    With an assumed equatorial radius of 71,400 km at a pres-
sure of 1 bar, Eq. (6) can be used to derive the polar radius
at the same level from the Pioneer 10 values of J_2 and J_4. The
result is b = 66,850 km, and the empirical value of K from Eq.
(14) is K = 1.62 (+0.5, -0.3) Mbar (g-cm^{-3})$^{-2}$, where the un-
symmetrical uncertainty is determined almost entirely by the
uncertainty of $\pm$ 0.00015 in J_4. The density at the sounding
level from Eq. (11) is 0.26 g-cm^{-3}, and the pressure from
Eq. (13) is about 110 kbar.

    The empirical determination of physical conditions in the
outer envelope of Jupiter has been interpreted in terms of a
family of theoretical pressure-density relations for a hydrogen-helium mixture in the 0-200 kbar range.[1] The results of
this comparison have yielded the somewhat surprising conclusion that the gravity data are more sensitive to assumptions
on temperature in the outer envelope than on the chemical composition. A series of adiabats with starting temperatures at
1 bar ranging from 200 to 340 K are shown in Fig. 1 for an
assumed composition of 73% hydrogen and 27% helium by mass.
These curves would not differ very much for other hydrogen-to-
helium ratios. The empirical determination of pressure and
density also is plotted in Fig. 1, along with two isotherms
that give some idea of the prevailing temperatures at the
sounding level for the assumed composition.

    The conclusion from Fig. 1 is that the temperature of
material is 250 $\pm$ 40 K at point where the atmospheric pressure
is 1 bar. This temperature is reasonably consistent with
results from the infrared radiometer on Pioneer 10[15] but con-
flicts with results from the S-band radio occultation experiment that obtained a detailed temperature profile for the
Jupiter atmosphere.[16] The temperature gradient from the
occultation experiment is generally superadiabatic; the temperature increases with depth at a very high rate until at a
pressure of 1 bar we would expect a temperature well in excess
of 300 K, a value seemingly ruled out by the gravity data.

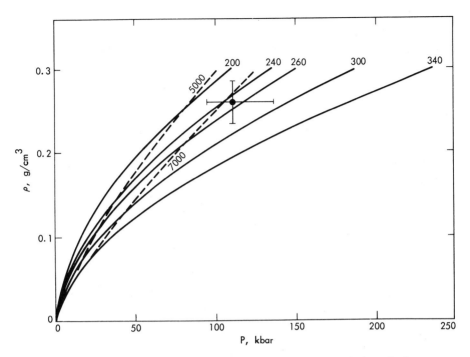

Fig. 1 <u>Solid lines</u>, adiabats for solar composition hydrogen
and helium. Liquefaction occurs above a density of 0.01 g
cm^{-3}. The starting temperature at one bar is given in degrees
K above each adiabat. <u>Dashed curves</u>, 5000K and 7000K isotherms, respectively. Error box is obtained from the observed
value of J_4.

However, it should be remembered that the gravity data provide
a technique for probing the outer envelope of Jupiter to a
depth that is not accessible to the occultation experiment.
On the other hand, they do not determine atmospheric conditions directly. Eventually, the most satisfactory model for
the outer envelope of Jupiter will be obtained by combining
results from the infrared, radio occultation, and gravity
experiments. Inconsistencies with the radio occultation data
may be resolved by additional analysis and by the Pioneer 11
flyby, but, for really satisfactory consistency, it may be
necessary to wait for Mariner flybys in 1979 or perhaps to a
time when probes will enter the atmosphere itself. The con-
straint imposed by the gravity data on the structure of Jupiter
is expected to improve with the analysis of the Pioneer 11 data
and with the construction of complete gravitational models of
the interior of the planet.

REFERENCES

1
Anderson, J. D., Hubbard, W. B., and Slattery, W. L.,
"Structure of the Jovian Envelope from Pioneer 10 Gravity
Data," Astrophysical Journal Letters, Vol. 193, November
1974, pp. L149-L150.

2
Anderson, J. D., Null, G. W., and Wong, S. K., "Gravity
Results from Pioneer 10 Doppler Data," Journal of Geo-
physical Research, Vol. 79, Sept. 1974, pp 3661-3664.

3
Brouwer, D. and Clemence, G. M., "Orbits and Masses of
Planets and Satellites," Planets and Satellites, edited by
G. P. Kuiper and B. M. Middlehurst, University of Chicago
Press, Chicago, Ill., 1961, pp 31-94.

4
Whitaker, E. A. and Greenberg, R. J., "Eccentricity and
Inclination of Miranda's Orbit," Monthly Notices of the
Royal Astronomical Society, Vol 165, 1973, p 15.

5
Garcia, H. A., "The Mass and Figure of Saturn by Photo-
graphic Astrometry of its Satellites," Astronomical
Journal, Vol. 77, 1972, pp 684-691.

6
Zharkov, V. N. and Trubitsyn, V. P., "Theory of the Figure
of Rotating Planets in Hydrostatic Equilibrium - A Third
Approximation," Soviet Physics-Astronomy, Vol. 13, May-
June 1970, pp 981-988.

7
Gulkis, S. and Poynter, R., "Thermal Radio Emission from
Jupiter and Saturn," Physics of the Earth as a Planet.
Interiors, Vol. 6, 1972, p 36.

8
Chandrasekhar, S., An Introduction to the Study of Stellar
Structure, University of Chicago Press, Chicago, Ill.,
1939, p. 40.

9
Podolak, M. and Cameron, A. G. W., "Models of the Giant
Planets," Icarus, Vol. 22, June 1974, pp. 123-148.

10
De Marcus, W. C., "The Constitution of Jupiter and Saturn,"
Astronomical Journal, Vol. 63, 1958 p. 2.

11
Peebles, P. J. E., "The Structure and Composition of
Jupiter and Saturn," Astrophysical Journal, Vol. 140,
p. 328.

[12] Hubbard, W. B., "Thermal Models of Jupiter and Saturn," Astrophysical Journal, Vol. 155, Pt. 1, Jan. 1969, pp. 333-344.

[13] Hubbard, W. B., "Structure of Jupiter: Chemical Composition, Contraction, and Rotation," Astrophysical Journal, Vol. 162, Pt. 1, Nov. 1970, pp. 687-697.

[14] Hubbard, W. B., "Inversion of Gravity Data for Giant Planets," Icarus, Vol. 21, 1974, p. 157.

[15] Chase, S. C., Ruiz, R. D., Munch, G., Neugebauer, G., Schroeder, M., and Trafton, L. M., Science, Vol. 183, 1974, p. 315.

[16] Kliore, A., Cain, D. L., Fjeldbo, G., Seidel, B. L., and Rasool, S. I., "Preliminary Results on the Atmospheres of Io and Jupiter from the Pioneer 10 S-band Occultation Experiment," Science, Vol. 183, 1974, p. 323.

ATMOSPHERES OF OUTER
PLANET SATELLITES

R. W. Carlson
University of Southern California, Los Angeles, Calif.

Abstract

Only three of the many satellites in the outer solar system are now known or thought to possess atmospheres: Io, Ganymede, and Titan, and the physical properties of these atmospheres are briefly reviewed here. Evidence for an atmosphere around Io (Jupiter-I) is found in recent optical observations of sodium and hydrogen atomic resonance emissions associated with the satellite. These emissions are found to originate from a volume which is much greater than Io itself, forming a partial toroid around Jupiter. Suggestive evidence for an atmosphere on Ganymede (Jupiter-III) is found in stellar occultation measurements. In light of spectroscopic limits on CH_4 and NH_3, other possible atmospheric constituents are considered along with their production and loss mechanisms. These brief considerations suggest that A, O_2, N_2, and Ne are possible candidates for an atmosphere on Ganymede. Titan (Saturn-VI) has been long known to possess a CH_4 atmosphere, but recent work indicates that the amount present is greater than had originally been estimated. In addition, evidence for molecular hydrogen has also been found. The high thermal-infrared brightness temperatures of this satellite have also received much attention,

It is a pleasure to acknowledge many interesting discussions with T. V. Johnson, D. L. Judge, D. L. Matson, and T. R. McDonough. Portions of this work were supported under NASA Grant NAS-2-6558 with the Ames Research Center.

*Research Staff Physicist, Department of Physics.

and have been interpreted as a high altitude inversion layer
or a greenhouse effect.

Introduction

As early as 1921, in discussing the escape of gases
from planetary atmospheres, Sir James Jeans[1] wrote that
"an atmosphere has been observed on Titan" and then goes on
to mention "the suspected atmospheres on two of Jupiter's
satellites." These would seem very prescient remarks indeed, since the only outer planet satellites which are presently thought to possess atmospheres are Titan and two of
the Galilean satellites of Jupiter, and the evidence for these
was not found for twenty years or more after Jeans wrote
those thoughts. It should not be surprising, however, that
some of the satellites possess atmospheres. They (and other
outer solar system satellites) are comparable to, or exceed,
the planet Mercury in size and although less massive they are
of sufficiently low temperature that thermal evaporation
(Jeans' escape) is greatly reduced. What is surprising are
the details of actual atmospheres, which represent extremes
as great as can be found among the planets. On the one hand,
Titan exhibits a very thick and relatively permanent molecular atmosphere, while in contrast Io is seen to possess a
tenuous and short-lived atmosphere composed of atomic hydrogen and metal atoms. The major portion of our body of
knowledge concerning these atmospheres has been established
in only the past few years, and the field is rapidly developing.
Many of the present activities are directed toward specific
atmospheric questions, e. g. , composition, temperature
photochemistry, and escape processes, but it is also recognized that these studies have even greater implication since
the atmospheres are related to surface and interior compositions and the environment provided by the central planet.
Their continued study will aid in understanding these environments and the physical-chemical history of the satellites and
will surely be among the major scientific objectives in future
missions to the giant planets.

The purpose of this work is to condense and summarize
for the non-specialist, recent developments in studies of
these atmospheres and to offer some speculation on what fu-

ture developments may offer. The Galilean satellites have been discussed by Cruikshank[2] while Titan's atmosphere has received enough attention to warrant a volume of its own.[3] The general properties of the physical satellites has been reviewed by Morrison and Cruikshank.[4] An extensive summary of the outer solar system by Newburn and Gulkis[5] is also of interest. A summary of some of the pertinent data concerning these objects is given in Table I.

Io

Post-Eclipse Brightening

The first evidence, still controversial, for a rarefied atmosphere on Io is found in its apparent anomalous photometric behavior following passage through the shadow of Jupiter - post eclipse brightening. This phenomena, first observed by Binder and Cruikshank[7], shows an excess brightness for the satellite of~ 0.1 mag (~ 10%) immediately following eclipse which decays during the succeeding 10-20 minutes. The phenomena is not seen for the other satellites nor is it observed before eclipse.

Binder and Cruikshank[7] suggested that a condensable atmosphere produced this differential brightness by forming a surface layer or haze of brighter material (frost or snow) by condensation during eclipse cooling. It then sublimes back into the atmosphere in a short period following eclipse. Lewis[8] presented physical-chemical models of the larger outer satellites and discussed the atmospheric implications of these models. For Io, he preferred NH_3 or an inert gas,

Table I Physical Properties

	Mass (10^{26}gm)	Radius (km)	Density (gm/cc)	Escape Velocity (km/sec)
Io (J-I)	0.910	1820	3.50	2.58
Ganymede (J-III)	1.490	2635	1.95	2.75
Titan (S-VI)*	1.401	2500	2.14	2.73
		(2900)	(1.37)	(2.53)

*Values in parentheses are based on recent lunar occultation measurements.[6]

rather than CH_4, as the suspected condensate and estimated the atmospheric abundance by two methods. He first noted that it would require of order 1 mg/cm^2(or~ 1 cm-atm) to produce the differential reflectivity. Secondly, the amount of ice that can be sublimed by absorption of the available solar energy in the 15 min post-eclipse period (assuming an albedo of 0.8) is roughly the same (~ 0.8 cm-atm). At the temperature of Io, ~ 140ºK, this abundance would produce a surface pressure of~ 1 x 10^{-7} bar, comfortably below the upper limit of 3 x 10^{-5} bar set by the NH_3 vapor pressure. The agreement between these estimates supports Binder and Cruikshank's original suggestion as a viable hypothesis.

A problem posed by the post-eclipse brightness phenomena is its sporadic nature. Many eclipse reappearances have been investigated by different observers using a variety of techniques, yielding both positive and negative results, the latter casting some doubt as to the reality of the effect. Unfortunately, very few events have been observed simultaneously by several observers and it is therefore difficult to differentiate between mere instrumental effects caused by scattered light from Jupiter and a truly satellite-related phenomena which is intermittent in occurrence.

The erratic nature of the effect prompted Fallon and Murphy[9] to suggest the possibility of transient atmospheres, perhaps due to irregular outgassing. An alternative idea was advanced by Cruikshank and Murphy[10], who argued that a strong temperature variation in vapor pressures, and the temperature differences between perhelion and aphelion, could produce an effect varying with the Jovian anomalistic year since considerably more gas (of unspecified composition) would be available at perhelion where the mean temperature is higher. A seasonal effect was proposed by Sinton[11], arguing that during the solstices non-illuminated polar regions would act as a cold trap, freezing out much of the postulated NH_3 atmosphere, but sufficient atmosphere is present at the equinoxes to produce the post eclipse brightening anomaly. He presented a rather detailed model of the Ionian atmosphere, consistent with then existing measurements and containing~ 0.5 cm-atm of NH_3, heated to~ 245ºK and perhaps including as much as 4 cm-atm of molecular nitrogen.

Recent observations[12] using instruments which are less
sensitive to scattered light have given negative results.
These data are of particular interest since they were obtained during a period when both Sinton's and Cruikshank and
Murphy's hypotheses predicted positive results.

Spectroscopic and Occultation Limits

Various spectroscopic studies have been performed on Io
and the remaining Galilean satellites, resulting in upper limits for the presence of certain gases. Kuiper[13] placed limits
of 200 cm-atm and 40 cm-atm respectively for CH_3 and NH_3.
Owen[14] extended these measurements by photographing the
infrared spectrum; based on the absence of the strong 8873 Å
band he placed a limit of 100 cm-atm of methane. Recently,
Fink et al.[15] using a Michelson interferometer, looked for
features of CH_4 and NH_3 in the 2.3 μ region. Finding none,
they were able to place limits of 0.5 cm-atm for both NH_3
and CH_4 which corresponds to 6×10^{-8} bar partial pressures.

Another technique - stellar occultations - can be used
to discern the presence of an atmosphere (or the lack thereof).
If an object with sufficient atmospheric density passes in
front of a star, refraction will bend the light grazing the atmosphere toward the center of the shadow (if the atmosphere
is normally dispersive). This bending produces a gradual
shadow boundary whereas an abrupt boundary would be found
in the absence of an atmosphere. The first photoelectric observations of an occultation were performed by Baum and
Code[16] who observed an occultation by Jupiter. In May 1971,
Io occulted the C component of β Scorpii and the event was
observed by several groups[17-19]. All of the observations
showed a sharp light curve, within limits set by instrumental
time response, diffraction, and the finite stellar diameter.
Smith and Smith[18] placed limits on the refractivity of the gas
at the surface and therefore the corresponding number dens-
ity. For N_2, CH_4, and H_2 these limits are 6×10^{12}, 9×10^{12}
and 3×10^{15} molecules/cm^3 respectively. Assuming a temperature of $100°K$, the surface pressure limits are 9×10^{-8}
and 1.3×10^{-7} bar for N_2 and CH_4, corresponding to column
abundance limits of 0.4 and 1.3 cm-atm. Limiting NH_3 abundance would be of the same order, and may be consistent

with the atmospheric abundances suggested by Lewis[8] and Sinton[11] particularly since the occultation pressure and abundance limits vary as $T^{5/2}$.

Sodium Emissions

The above discussion indicates that evidence for an atmosphere on Io is largely negative: rather severe upper limits are placed by the occultation and spectroscopic measurements and the post eclipse brightening anomaly remains unconfirmed. Definite and conclusive proof of atmospheric phenomena surrounding Io, far different than ever would have been expected, was recently discovered by Brown[20,21]. In his first planetary observational work, he obtained spectra of the Galilean satellites which showed sodium D line emission features from Io. These results were first reported in 1973 and quickly confirmed by other observers.[22-24] A spectrum obtained with a Wampler type coude scanner of the JPL Table Mountain Observatory is shown in Fig. 1. The spectral shift of the line due to the Doppler effect and Io's orbital velocity is clearly evident. The intensities observed by the various groups is variable (discussed below) with average intensity being many tens of thousands of Rayleighs.[25]

Trafton et al.[22] and Macy and Trafton[23] investigated the distribution of these emission features around Io, finding that the source is an extended region around the satellite extending as great as 50 Io radii from the satellite (in the orbital plane) and roughly 5 radii above the plane. The most intense region appears to be an area around Io whose radius is approximately twice the radius of Io.

The D lines, the resonance lines of neutral sodium, are well known features in the terrestial airglow, occurring in the day, twilight, and nightglow, and in aurora, the source of the sodium atoms being meteoritic with perhaps some contribution from oceanic salt. The excitation mechanisms for the earth's sodium emissions are resonance scattering (day and twilight), chemical excitation, and energy transfer from vibrationally excited N_2 (aurora). Reviews and discussions o of the terrestial Na problem are described by Hunten[26], Chamberlain[27] and Bates[28].

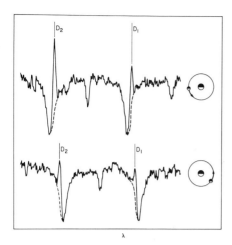

Fig. 1. Coudé spectra of sodium emissions from Io obtained at the Table Mountain Observatory by J. T. Bergstrahl, D. L. Matson and T. V. Johnson (Reference 24). The orbital position of the satellite is indicated on the right. The sun is to the bottom of the page.

The problems posed by the existence of metal emission features from Io can be classified as (1) the source of the Na and the mechanism by which it introduced into the extended atmosphere, (2) the excitation mechanism for the observed emission, (3) the eventual fate of the atoms after they escape the influence of the satellite, and (4) what kinetic, ionospheric, and photochemical interactions occur during the interval between production and loss?

An answer to the first question is provided by the studies of the Jet Propulsion Laboratory group, Fanale et al.[29], who suggest that the surface composition of Io involves salt deposits enhanced in sodium, while Matson et al.[30] argue that sodium atoms are liberated from the surface and injected into the atmosphere by sputtering processes.

In the model of Fanale et al.[29], Io and the remaining Galilean satellites were formed from chondritic material and H_2O with the relative proportions differing due to the influence of Jupiter acting as a significant source of heat in the early stages of formation. Subsequent radioactive heating in the interior melted the ice and bound water which then percolated to the surface and evaporated. As this water passed through the interior it became saturated with salts, carrying them to the surface and depositing them upon evaporation. The hypothesis is supported by several experimental studies

including the spectral reflectivity of Io and laboratory experiments with carbonaceous chondritic material.

Matson et al.[30] argue that the sodium is removed from the
surface by particle impact - sputtering. One source of the
impacting particles could be the energetic magnetospheric
ions, another source may involve the plasma sheaths thought
to exist around Io[31]. The induced EMF across Io in the moving Jovian magnetic field may produce plasma sheaths around
the satellite which in turn can develop large electrical potentials between the plasma and areas on the surface of Io. If
ions are produced in these regions, they may be accelerated
into the surface with energies of perhaps several hundred Kev.
The source of these ions could be previously sputtered atoms
which are photoionized or directly ionized in the sputtering
process.

Two other suggestions have been advanced to explain the
presence of Na. McElroy et al.[32] suggest that Na and other
metal atoms are present in solution in ammonia ice on the
surface but they do not offer a mechanism for transfer to the
atmosphere. It should be noted that the reflection spectrum
of Io does not show any ice absorption bands in contrast to
the remaining Galilean satellites. Sill[33] suggested that the
sodium cloud originates from meteoritic material swept up
by Jupiter, followed by decomposition by the energetic trapped radiation belt particles. This model does not explain the
unique association of the sodium cloud with Io, however.

A mechanism for producing the observed Na-D emissions
was suggested by McElroy et al.[32] as energy transfer from
vibrationally excited N_2, a process that is thought to occur in
terrestial aurorae. The initial vibrational excitation was
thought to be produced by aurorae - like phenomena at Io, the
N_2 being present as a photolysis product of their assumed
NH_3 atmosphere. Two factors guided this choice of excitation mechanism. First, the observational data available at
that time seemed to show that the emission was sporadic (as
are aurorae) and second, the emitting region was thought to
be highly localized which implied higher surface brightnesses
than could be supplied by resonance scattering of sunlight.

Further observational data by Trafton et al.[22] showed that the emission arose from a more extended region than originally was supposed meaning a lower surface brightness and suggesting to Matson et al.[30] and Trafton et al.[22] that resonance scattering was the dominant excitation process.

Synoptic measurements by Bergstrahl et al.[24] at the Table Mountain Observatory showed that the emission was not sporadic, but varied smoothly with Io's rotational phase as seen from the sun and that intensities at the same orbital phase were quite repeatable over time scales covering many revolutions. The rotational or orbital variation is due to the Doppler shift of the atoms relative to the sun which modulates the emission intensity because of variations in intensity of the solar emission over the Fraunhofer line profile, the Swings effect, so well known in comets. The Table Mountain data agree well with theoretical predictions of the orbital phase variation when the solar profiles are taken into account and prove that resonance scattering is the dominant source. The long term observations demonstrate that the densities of Na are roughly constant, arguing against sporadic auroral phenomena.

Observations by Trafton et al.[22] and Macy and Trafton[23] show that sodium is not confined to the immediate vicinity of Io, but forms an extended cloud far outside the gravitational influence of Io. The atoms must therefore escape from Io, whose escape velocity is 2.5 km/sec, and the subsequent dynamics are determined largely by the gravitational potential of Jupiter. After escape, the atoms will orbit Jupiter in Keplerian trajectories as they do not in general possess enough energy to escape the vicinity of the planet, and the resulting density distribution would tend to form a cylindrically symmetric toroidal distribution unless the lifetime is limited. McDonough and Brice[34,35] were the first to point out the possible existence of gaseous toroids around the major planets. One process which will limit the lifetime is photoionization. However, the lifetime of Na atoms against photoionization at the orbit of Jupiter is $\sim 1.5 \times 10^6$ sec, much greater than Io's orbital period (1.5×10^5 sec) and the torus would be expected to be more complete than is observed.[22-23] In addition to photoionization, Macy and Trafton[23] investiga-

ted ionization by ions in the plasmasphere of Jupiter, finding
this mechanism an inadequate explanation owing to the low
charge exchange cross section for non-resonant systems. It
is suggested here that electron impact ionization by thermal
electrons in the Jovian plasmasphere is the dominant loss
mechanisms. Using the ionization rate coefficients of Lotz,[36]
a plasma energy of 4 ev (see Intriligator and Wolfe[37]), and
plasma densities of $200/cm^3$ obtained by Carlson and Judge[38]
and McDonough[39], the lifetime of Na atoms is found to be
$\sim 1.1 \times 10^5$ sec, in reasonable agreement with the lifetime
estimated by Macy and Trafton[23] based on their observations
of the cloud geometry.

Interactions with other possible atmospheric constituents
by the sodium atoms during the interval between production
and loss is unknown at present. Part of this uncertainty
is due to lack of knowledge of other components in the atmosphere, and further uncertainty lies in the poor state of knowledge concerning the gas phase chemistry of Na. It seems
possible that some other component is present but the total
amount of atmosphere must be small, or the exospheric
temperature must be very high, else the sodium atoms would
lose sufficient energy in collisions that escape would be impeded too greatly. Further observational work, coupled with
investigations of Na gaseous chemistry should delineate some
of the possibilities. Certain of the possible candidates as
atmospheric components (e.g. N_2, H_2, O, A, Mg, N, C, Si)
are most readily observed in the extreme ultraviolet and may
not be properly investigated until suitable instrumentation
is available for long term study on a Jupiter orbiting mission.

Hydrogen Emissions

A second component, atomic hydrogen, was detected in
an extended cloud around Io through ultraviolet photometric
measurements of the HI Lyman - α line reported by Judge
and Carlson[40] and Carlson and Judge[38,41]. The excitation
mechanism is similar to the sodium case: resonance scattering of the incident solar flux, and the distribution of atoms
is also qualitatively similar, forming an incomplete torus

around Jupiter approximately centered at Io. While it is apparent that Io is responsible for the observed distribution, it is less clear what the actual mechanism is that produces the atomic hydrogen. McElroy et al.[32] suggests that copious amounts of hydrogen will be produced in the photolysis of the hypothetical NH_3 atmosphere and will readily escape the planet. An alternative hypothesis[38, 39] suggests that magnetospheric and plasmaspheric protons are neutralized in the atmosphere and surface of Io, followed by escape into the torus.

In addition to being a possible source mechanism, the Jovian plasma protons are very likely involved in the H atom destruction processes. The charge exchange cross section between protons and neutral hydrogen atoms is very large,[42] being an exact energy resonance condition. If the slowly moving H atoms charge exchange with protons co-rotating with the Jovian magnetic field, the resultant will be slow ions and energetic neutrals which possess enough velocity to escape Jupiter, thereby depleting the torus. Estimates of the lifetime of H atoms in the torus, based on the cloud geometry, yield values of $\sim 2 \times 10^5$ sec, which imply plasma densities of 200 cm^{-3} or somewhat greater (Carlson and Judge[38, 40]) and are in reasonable agreement with the theoretical discussion of McDonough[39].

Ionosphere

Radio occultation measurements have been very valuable in studying the ionospheres and atmospheres of the terrestrial planets, and the same technique was recently applied to the atmosphere of Io. Kliore et al.[43] detected an ionosphere on this satellite with a measured peak electron density of $\sim 6 \times 10^4$ cm^{-3} on the day side and $\sim 10^4$ cm^{-3} on the night side. The day side plasma scale height was found to be 220 km. They note that if the Mars atmosphere is a close analogy, then atmospheric densities of 10^{10}-10^{12} cm^{-3} would be found at the surface where the pressure would be 10^{-8}-10^{-10} bar. This corresponds to column abundances of 0.003 - 0.3 cm-atm and below the aforementioned stellar occultation limits. These investigators point out that one is probably seeing a rather different type of ionosphere, since it is more fully ionized than one would expect at this solar distance, and since the ionosphere probably extends down to the surface.

McElroy and Yung[44] investigated the ionospheric properties of several model atmospheres, including both purely
molecular atmospheres and those containing sodium. They
found that ionization rates were too slow, and recombination
too fast for a molecular atmosphere to exhibit the observed
ionosphere. Inclusion of atomic sodium which is ionized
much more rapidly and recombines more slowly brings the
calculated densities closer to reality although complications
such as diurnal atmospheric abundances, vertical motions,
plasma interactions, and corpuscular ionization may enter
into the problem and are difficult to distinguish with our
present limited state of knowledge.

Further observations are clearly desirable in order to
fully describe the spatial and temporal ionospheric variations and allow a more complete theoretical description.
Since occultation measurements in the outer solar system
are limited to measurements near the terminator, a complete description may not be forthcoming until plasma probes
(and hopefully mass spectrometers) investigate the ionosphere close to the satellite on an orbiter mission.

Ganymede

The present state of knowledge concerning an atmosphere
is suggestive but not conclusive, similar to the case of Io
prior to the discovery of sodium and hydrogen emissions and
the Ionian ionosphere. Whereas a stellar occultation provided an upper limit to the atmosphere of Io, a similar event
with Ganymede did show evidence of an atmosphere. In the
following section we discuss these results, followed by
speculation on the possible atmospheric constituents.

Occultation Results

On 7 June 1972, Ganymede occulted the 8th magnitude star
SAO 186800 and observations were made in three locations:
Kodaikanal, India; Lembang (Java), Indonesia; and Darwin,
Australia and reported by Carlson et al.[45]. Unfortunately,
the occultation path occurred more northerly than had been

predicted and the Australian site was ~ 50 km too far south to observe the event. The measurement was a difficult one as the difference in magnitudes resulted in an intensity drop of only $\sim 5\%$. Nevertheless, the data were of sufficient quality to determine the radius and suggest, on the lack of abrupt intensity changes in the signal, the presence of an atmosphere. An approximate lower limit to the surface pressure was placed at 10^{-3}mb, which would correspond to ~ 5 cm-atm for a constituent with a mean molecular mass of ~ 30. The major component cannot be methane or ammonia, since Fink et al.[15] placed upper limits on these molecules in order of magnitude less than densities suggested by the occultation.

Discussion of Possible Atmospheric Constituents

The possible presence of an atmosphere other than the two attractive candidates, NH_3 and CH_4, poses an interesting problem, particularly since the Jovian environment offers unique processes for both producing and depleting an atmosphere. We begin by discussing possible sources for an atmosphere, followed by remarks on loss processes.

The first source one might consider is outgassing from the interior of the satellite. As is well known, the rare gases He and Ar present in the terrestial atmosphere arise as radioactive decay products from uranium, thorium, and potassium (K^{40}, which produces A^{40} through K capture decay). Helium will escape so readily from Ganymede that it cannot be the specie observed in the occultation. An approximate upper limit to the argon content in the atmosphere is readily estimated assuming solar composition for elements other than hydrogen and helium and assuming the mass contribution by H is as H_2O ice. Using the isotopic abundance of K^{40}, the half life of 1.3×10^9 years, and the K capture branching ratio of 11%, one finds a present day production rate of $\sim 10^5$ A^{40} atoms/cm^2 sec and a total of $\sim 6 \times 10^{22}$ (20m-atm) produced over geologic time. It is unlikely that all of this gas has reached the surface and even more unlikely, as discussed below, that all of it would be permanently retained in the atmosphere.

In analogy with the terrestial atmosphere, N_2 might be considered as a possible contribution from outgassing, but it is impossible to make any quantitative estimates. Presumably, most of the nitrogen in the solar nebula that formed the outer solar system was in the form of NH_3, and the ammonia present at Ganymede is to be found in solution in a liquid water mantle[8], However, N_2 could be formed as a photolysis product of any small amount of NH_3 which finds its way to the surface and into the atmosphere.

The surface itself, composed of H_2O ice, could be a source of atmospheric molecules through the process of sputtering, as is found for Io. In the case of Ganymede, sputtering would probably produce, in addition to H_2O molecules, H, OH, and O, the latter atoms possibly recombining to produce an atmosphere of O_2. The production would be self-limiting, however, since a modest atmosphere would shield the surface from the particles responsible for the sputtering. (The range of a 1 MeV proton is ~2 cm-atm).

The magnetospheric particles themselves could be a source for the atmosphere, whether or not they impact the surface. Incident particles will be stopped at the surface or in the atmosphere, recombine, and thereby contribute to the atmosphere. The major component of the magnetospheric plasma, protons, will immediately escape as hydrogen atoms, but ions with greater mass may also be present in the plasma. If one assumes that the source of the Jovian magnetosphere is the solar wind, and that solar abundances are maintained, then elements worthy of consideration are O, C, N, and Ne. These atoms are both massive enough to not be readily lost through thermal evaporation and sufficiently abundant that they may significantly contribute to the atmosphere. Assuming a total plasma flux of 10^6 cm^{-2} sec^{-1}, with solar abundances, one might expect ~10^3, 5×10^2, 10^2 and 10^2 cm^{-2} sec^{-1} for O, C, N, and Ne respectively. Atomic oxygen and nitrogen can associate to form relatively inert atmospheric molecules by surface catalysis or by atmospheric three body reactions.

The above source mechanisms must be weighed against possible loss processes in order to evaluate their possible contribution. One obvious escape mechanism is Jeans' escape or thermal evaporation; this mechanism is inefficient for the heavier atoms and molecules such as Ne, N_2, O_2 and A. If Jeans' escape is the only mechanism by which particles escape (an unlikely situation) then one can estimate the effectiveness of the sources discussed above in terms of a critical temperature for the outer atmosphere. This critical temperature is defined for each molecule such that the escape rate just equals the influx rate. If the temperature is below the critical temperature, then the net accumulation rate (influx less loss) can result in the accumulation of an atmosphere over geologic time. In Table 2 we summarize the possible influx rates and corresponding critical temperatures. If the exospheric temperature is comparable to the surface temperature (~ 140°K) and Jeans escape were the only loss mechanism, then it is possible that significant amounts of A, O_2, N_2, and Ne could accumulate to form a modest atmosphere on Ganymede.

Jeans' escape represents only a lower limit to the total escape rate; there are other mechanisms which are potentially much more rapid. One other means of escape is through molecular dissociation processes such as photodissociation and dissociative recombination, where the fragments are produced with sufficient kinetic energy to escape gravitational attraction as suggested by Brinkmann[46] for the atmosphere of Mars.

Table 2   Critical Temperature for Accumulation
of an Atmosphere

Specie	Assumed Flux ($cm^{-2}sec^{-1}$)	Critical Temperature ($^\circ$K)
A	10^5	500
O_2	10^3	200
N_2	10^2	300
Ne	10^2	200

A third class of escape mechanisms is brought about
through ionization and the magnetic environment of Jupiter.
Since Jupiter and its magnetic field rotate much faster than
the orbital period of Ganymede, a V x B field will be produced which is several orders of magnitude greater than the
gravitational force and can accelerate atmospheric ions away
from the planet into the magnetosphere. This is thought to
be the means by which the lunar atmosphere is depleted[47],
in the lunar case the velocity and magnetic field being properties of the solar wind rather than a planetary field. In the
absence of natural magnetic shielding (or diamagnetic behavior as discussed below), potentially every ion created by solar ultraviolet photoionization could be lost by this process.
Since the total solar ultraviolet which can ionize an N_2 atmosphere (for example) at the orbit of Jupiter is ~10^9 photons
cm^{-2} sec^{-1}, and ionizations yields are close to unity, the
V x B acceleration could play a dominant role in the loss of
a gaseous atmosphere.

One can examine the efficiency with which the induced
electric field removes ions by finding the time it takes an
ion moving under the induced V x B field to leave the atmosphere and comparing it to the recombination time. If the
recombination time is large compared to the time necessary
to remove the ion, then escape could be very efficient; if the
lifetime of an ion and the time interval during which the
V x B potential can act is shorter then this loss mechanism
could be impeded.

The recombination time can be evaluated as $\tau_r = 1/\alpha n(e)$
where α is the recombination rate coefficient and n(e) the
electron density. This density is estimated here by assum-
ing local equilibrium: $\alpha \cdot n^2(e) = n_o/\tau_p$ where n_o is the neutral density and τ_p the neutral lifetime against photoionization. The recombination lifetime is thus $\tau_r = \sqrt{\tau_p}/\alpha n_o$.
The time necessary to remove an ion τ_{loss} may be approximated by the time required to move a plasma scale height
(twice the neutral scale height, H), with an average drift
velocity v_d, the latter quantity being related to the induced
electric field through the ionic mobility K: $v_d = KE$, resulting in $\tau_{loss} = 2H/KE$.

The induced electric field at Ganymede is $\sim 10^{-4}$ v/cm, and ionic mobilities (at STP) are $K \sim 1.5\text{-}2$ cm^2/volt-sec, resulting in $\tau_{loss} \sim 10$ sec for an assumed scale height of 25 km.

For the atoms of interest, the peak ionization rate will occur at densities $n_o \sim 2 \times 10^{10}$ cm^{-3} with ionization lifetimes of 5×10^7 sec. For atoms as Ne and A, recombination occurs radiatively with $\alpha \simeq 10^{-11}$ cm^3/sec giving a recombination lifetime of $\tau_r \simeq 10^4$ sec. Consequently, an atmosphere of Ne or A could be readily diminished by the magnetic field of Jupiter. Molecular recombination processes (dissociative recombination) are much faster than radiative processes so the situation for a molecular atmosphere is more favorable. For $N_2, \alpha \simeq 3 \times 10^{-7}$ cm^3 sec^{-1} resulting in a recombination lifetime of 100 sec. We therefore conclude that a molecular atmosphere is less likely to be swept away by magnetic fields, but nevertheless high loss rates may be suffered from this process unless some sort of shielding is available.

A permanent magnetic field could provide such a shielding mechanism, as the earth's atmosphere is shielded from the solar wind (except during geomagnetic reversals). While the possibility of a satellite magnetic field is not ruled out by observations, it seems somewhat unlikely since Ganymede rotates so slowly it would be difficult to generate an internal dynamo.

Another, more likely, possibility is an ionospheric interaction generated by the Lorentz force. One can discuss such an interaction from several points of view. One description is in terms of a polarization field which opposes the V x B force. Since electrons and ions would be accelerated toward, and removed from, opposite faces, one can imagine excess "surface" charges (residing in the outer ionosphere) which produces an opposing electric field and retards ion and electron loss from the atmosphere. Another process is described in terms of the magnetic field produced by ionospheric currents as was considered by Dessler[48] for the interaction of Mars with the solar wind plasma and magnetic field. He argues that if the conductivity of a body is high enough so

that the time required for magnetic diffusion is longer than
the time required for the solar wind to sweep past the body,
then the interaction will produce currents and a magnetic
field which decreases the V x B field in the body. If the Jovian magnetic field and the co-rotating thermal plasma can be
treated analogously, and Ganymede possesses an ionosphere
of sufficient conductivity, then it is possible that such an interaction might occur, reducing the loss rate and producing
a very interesting magnetic and plasma structure in the vicinity of Ganymede which would be observable by appropriate
magnetometers and plasma probes.

The above comments are necessarily incomplete, but they
indicate that if Ganymede does in fact possess an atmosphere,
then it could be the product of some very interesting physical
processes, both with regards to its production and loss
mechanisms. Clearly, further observations are desirable.

Titan

Titan was the first satellite to be found to possess an atmosphere when, in 1944, Kuiper[49] found CH_4 absorption
bands at 6190 Å and 7260 Å. He noted that the amount of methane was comparable to, but somewhat less than, that observed on Jupiter and Saturn. A later estimate was given as
200 m-atm[13]. Until recently, this value was accepted, and
Titan was thought to have a rather tenuous atmosphere with a
surface pressure of~ 2mb. This picture has changed drastically in the past few years, primarily due to spectroscopic
observations and investigations in the thermal infrared. This
work, supported by photometric and polarization measurements, indicates that Titan has a very thick atmosphere of
surprising composition and with a most interesting thermal
profile. Sufficient interest has been generated that a Titan
workshop was held at the NASA Ames Research Center in
1973 under the chairmanship of D. M. Hunten[3].

Spectroscopic Studies

The spectroscopic work on Titan is primarily due to Trafton, who investigated the methane bands and found a factor of
ten more methane[50] than did Kuiper, assuming CH_4 to be the
major specie present. Since the spectral profile is deter-

mined by pressure broadening, the observations are compatible with less methane only with the addition of some other gas, and then only in much greater proportions, consequently the minimum total atmosphere present corresponds to a pure CH_4 atmosphere.

Trafton[51] also found evidence for a second component in the atmosphere - molecular hydrogen. This was a very unexpected result since the lighter gases would be expected to escape this satellite with such ease. Despite this potentially rapid loss mechanism, the spectral observations indicate the possibility of a large H_2 abundance, of order 5 km-atm, which implies that the loss must be inhibited by some mechanism and/or a large source of H_2 must be operating. McDonough and Brice[34, 35] attempted to resolve the H_2 loss problem by recycling thd gas in a torus around Saturn. They pointed out that atoms which escape from Titan do not possess sufficient energy to escape from the central planet, but orbit Saturn until lost by ionization or recaptured by a satellite. They estimated that the effective loss rate could be reduced by as much as two orders of magnitude by this recapture process. The effectiveness of this process has been questioned by Hunten[52] who argues that recapture will increase the coronal densities and the escape rate until the next escape flux is the same as would be found in the absence of recycling. Hunten[53] pointed out that the escape of H_2 could be inhibited by diffusion in the atmosphere, and suggested an atmosphere containing something like 50 km-atm of N_2 would reduce the escape rate by roughly an order of magnitude. The N_2 could be formed through photolysis of NH_3 which may be present in the atmosphere in small quantities.

Trafton[54] also has presented intriguing evidence for an additional component in the form of unidentified features in the 1.06 and 1.1 μ regions. Such features are found in the spectrum of Uranus, but not Saturn, and could be due to methane photolysis products or to isotopic methane. Too little is known about the spectroscopic properties of these molecules and it is hoped that laboratory experiments will be pursued, along with the observational aspects, in order to further our understanding of Titan and the outer planets.

Thermal Structure

The second area which has prompted much of the interest in Titan is the high brightness temperatures found in the thermal infrared. For a body with the albedo of Titan at the orbit of Saturn, the sunlit disc-averaged brightness temperature would be expected to be ~ 110°K in the absence of an atmosphere. A thick atmosphere as Titan possesses would moderate the diurnal temperature variation, and the observed temperature would be expected to approach ~ 84°K. However, the observed temperatures over much of the spectrum are much higher than this. For example, the recent measurements of Gillett et al[55]. show temperatures of 158°K at 8μ, decreasing to 128°K at 13μ. It was suggested earlier by Allen and Murdock[56], who found a temperature of 125°K in the 10-14 μ region, that a greenhouse effect was occurring. This hypothesis was developed by Sagan[57] and Pollack[58] and it appeared that such was the case for Titan.

An alternate explanation was proposed by Danielson et al[59]. which was developed as a consequence of explaining the observed low albedo in the ultraviolet. They noted that this low albedo implied absorption by high altitude aerosol particles. The absorbed energy, in being transferred to the gas, would increase its temperature, and be re-emitted in molecular transitions. Two likely transitions, which may be present in the measurements of Gillett et al[55]. are CH_4 at 7.7μ and C_2H_6 at 12.2μ. The photochemical calculations of Strobel[58] show that C_2H_6 is present in sufficient quantities to produce the 12 μ feature. The high altitude absorbing aerosols could be solid methane or particles composed of the photolysis products of methane-i. e. -photochemical smog. The latter process seems certain to happen (particularly to a resident of Los Angeles), and has been quantitatively estimated by Strobel[60], who finds that approximately 20% of the methane dissociation irreversibly produces higher hydrocarbons.

The choice between the two models (or a combination of both) should be forthcoming as infrared and microwave observations are extended. Recent observations by Briggs[61] at 8085 MHz (3. 7 cm), have been reported as a radio brightness temperature of 115° ± 40°. Since all of the proposed atmospheric constituents are transparent at this frequency, these microwave results probably refer to the surface temperature. Using the emissivity of ice, Briggs[60] finds a surface temperature of $135^\circ \pm 40^\circ K$ which tends to support the existence of a greenhouse effect, although he cautions that the results could also be consistent with temperatures as low as $80^\circ K$ as would occur in the absence of a greenhouse effect. The measurements of Low and Rieke[62] do not show structure in the 17 and 28μ region which would be expected from the H_2 pressure induced transitions; this seems to rule out a massive H_2 greenhouse effect. Low and Rieke suggest a weak greenhouse effect with surface temperatures of $80-90^\circ K$.

It is important to note that many of the above measurements were interpreted using then existing radius estimates for Titan, ~ 2500 km. However, a recent lunar occultation observed by Veverka[6] gives a much larger radius (2900 ± 200 km), which would significantly reduce the numerical values of brightness temperature, bringing them down to values which could be expected without a large greenhouse effect.

Future Possibilities

A stellar occultation would be of great value in determining the thermal profile of the atmosphere, although the resulting interpretations is composition dependent. The ingenuous suggestion of Brinkmann[63] has transformed the annoying "spikes" observed in stellar occultations, which result from atmospheric inhomogenities, into a powerful method for determining relative compositions. By observing an occultation in several wavelength bands, the wavelength dependence of the refractivity can be determined, allowing one to discern relative compositions. The slow wavelength variation of indices of refraction limits the method to relatively simple compositions of two (or perhaps three) components. Predictions of stellar occultations have been carried out by G. Taylor for the past twenty years. He is now using the SAO catalog for the brighter planets and satellites and has initiated a search for occultations by Titan of stars fainter than those listed in the SAO catalog. Unfortunately, no occultations appear in the offing for the very near future. On statistical grounds, one would expect ~ 2 useful occultations per year and an excellent event once every 5 years[64].

A spacecraft radio occultation would be extremely useful in determining the thermal picture of Titan, although the analysis is again composition dependent. A radio occultation, which obtains both the ionospheric profile and the lower atmosphere refractivity profile will also allow a choice between different model compositions, since the ionospheric profile would be quite different for a predominantly H_2 atmosphere than an N_2 atmosphere for example.

A radio occultation which probes the lower atmosphere and places a level for the solid surface would be of value in choosing between surface and interior models, since the methane atmosphere could arise from a CH_4 hydrate, solid or perhaps liquid CH_4, or even an H_2O-NH_3-CH_4 fluid with no real boundary, and the surface pressure (or absence of a surface) is different for these cases.

The composition of the atmosphere can be studied by suitable optical measurements on a fly by (or better an orbiting) spacecraft. Since the ratio of atmospheric scale height to planetary radius is comparable to the terrestial planets, a relatively closer approach is possible than for the major planets, allowing one to observe the atmosphere at the limb of the satellite without the overwhelming background from the bright disc. Some emissions of interest would be those of N_2, N_2^+, NH, CH, and CN, the latter radials being possible photochemical products of a CH_4-H_2-N_2 atmosphere. The background problem is not so serious in the ultraviolet where one can attempt to observe atomic resonance transitions (H, C, N), the Lyman and Werner bands of H_2, and the Birge-Hopfield bands of N_2.

Observations of the toroid would also be useful to study the escape problem of H_2, and to determine the relative amounts of H and H_2 escaping from Titan and the exospheric temperature. Taberie[65] has calculated the atmospheric density profile of atomic hydrogen and then computes the flux of atomic and molecular hydrogen into the torus. She finds that the flux ratio of H and H_2 varies from 10^{-6} for an atmosphere containing equal H_2 and CH_4 to 1.6×10^{-1} for a predominantly N_2 atmosphere. Consequently, measurement of the H_2 Lyman and Werner band emissions, in comparison to the HI Lyman-α line, will allow a determination of the rela-

tive torus abundances and provide useful information on the
atmosphere and its composition.

Concluding Remarks

The satellites of the major planets are a diverse collection of objects and currently of great interest. It is seen
that several of these bodies possess atmospheres which can
be related to surface or interior properties. Since atmospheres can be studied by remote optical sensing, without the
necessity of direct probes, much of the first information
about the history, interiors, and surfaces of the satellites
will be obtained through atmospheric studies. In the future
we can expect stellar and radio occultation measurements,
further optical observations, both ground based and in wavelength regions inaccessible from the ground, ionospheric
plasma probe experiments, outer atmosphere and toroid
mass spectra, and eventually probe missions directly into
the atmospheres.

These notes have been directed toward the atmospheres
of Io, Ganymede, and Titan. With regards to the remaining
satellites for which atmospheres have not been observed, we
close by recalling Kuiper's remark that the matter should
not be regarded as closed.

References

[1]Jeans, J.H., The Dynamical Theory of Gases, 3rd Ed.
Cambridge University Press, 1921, p.348.

[2]Cruikshank, D.P., "Atmospheres of Galilean Satellites",
presented at IAU Coll. No. 28, Planetary Satellites, Ithaca,
N.Y., to be published in the proceedings, Cornell Univ.
Press.

[3]Hunten, D.M., (Ed.), The Atmosphere of Titan, NASA SP-
340, Scientific and Technical Information Office, Washington,
D.C., 1974.

[4]Morrison, D. and Cruikshank, D.P., "Physical Properties
of the Natural Satellites", Space Science Reviews, Vol. 15,
1974, p.641

[5]Newburn, R. L. and Gulkis, S. , "A Survey of the Outer
Planets, Jupiter, Saturn, Uranus, Neptune, and Pluto, and
their Satellites", Space Science Reviews, Vol. 3, 1973,
p. 179.

[6]Elliot, J. L. , Veverka, J. , and Goguen, J. , paper presented at IAU Coll. No. 28, Planetary Satellites, Ithaca, N. Y. ,
1974.

[7]Binder, A. B. and Cruikshank, D. P. , "Evidence for an Atmosphere on Io", Icarus Vol. 3, 1964, p. 299.

[8]Lewis, J. S. , "Satellites of the Outer Planets: Their Physical and Chemical Structure", Icarus, Vol. 15, 1971, p. 174.

[9]Fallon, F. W. and Murphy, R. E. , "Absence of Post-Eclipse
Brightening of Io and Europa in 1970", Icarus, Vol. 15, 1971,
p. 492.

[10]Cruikshank, D. P. and Murphy, D. P. , "The Post-Eclipse
Brightening of Io", Icarus, Vol. 20, 1973, p. 7.

[11]Sinton, W. M. , "Does Io Have An Ammonia Atmosphere? ",
Icarus, Vol. 20, 1973, p. 284.

[12]Franz, O. G. and Millis, R. L. , "A Search for Post -
Eclipse Brightening of Io in 1973, II. ", Icarus, Vol. 23,
1974, p. 23.

[13]Kuiper, G. P. , in The Atmospheres of the Earth and
Planets, ed. by. G. P. Kuiper, University of Chicago Press,
1952.

[14]Owen, T. , "Saturn's Rings and the Satellites of Jupiter:
Interpretation of Infrared Spectra", Science, Vol. 149, 1965,
p. 974.

[15]Fink, U. , Dekkers, N. H. , and Larson, H. P. , "Infrared
Spectra of the Galilean Satellites of Jupiter", Astrophys. J.
(Lett.), Vol. 179, 1973, p. L155.

[16]Baum, W. A. and Code, A. D. , "A Photometric Observation of the Occultation of σ Arietis by Jupiter", Astron. J.,
Vol. 58, 1953, p. 108.

[17]Fallon, F. W. and Devinney, E. J. Jr. , "Observation of the
Occultation of β-Sco C by Io", Icarus, Vol. 17, 1972, p. 216.

[18]Smith, B.A. and Smith, S.A., "Upper Limits for an Atmosphere on Io", Icarus, Vol. 17, 1973, p. 218.

[19]Bartholdi, P. and Owen, F., "The Occultation of Beta Scorpii by Jupiter and Io, II, Io.", Astron. J., Vol. 77, 1972, p. 60.

[20]Brown, R.A., in Exploration of the Planetary System, Proceedings of IAU Copernicus Symposium (IAU Symposium No. 65), ed. by. A. Woszczyk and C. Iwaniszwiska, (in press).

[21]Brown, R.A. and Chaffee, "High Resolution Spectra of Sodium Emission from Io", Astrophys. J. (Lett.), Vol. 187, 1974, p. L125.

[22]Trafton, L., Parkinson, T., and Macy, W.M.Jr., "The Spatial Extent of Sodium Emission Around Io", Astrophys. J. (Lett.), Vol. 190, 1974, p. L85.

[23]Macy, W.M.Jr. and Trafton, L., "Io's Sodium Emission Cloud", To be published in Icarus, 1975.

[24]Bergstrahl, J.T., Matson, D.L., and Johnson, T.V., "Sodium D-Line Emission From Io: Synoptic Observations from the Table Mountain Observatory", Astrophys. J. (Lett.) Vol. 195, 1975, p. L1.

[25]The Rayleigh is a measure of surface brightness corresponding to $10^6/4\pi$ photons/sec-cm^2(column)-sterad. See Reference 27, p. 569.

[26]Hunten, D.M., "Spectroscopic Studies of the Twilight Airglow", Space Science Reviews, Vol. 6, 1967, p. 493.

[27]Chamberlain, J.W., Physics of the Aurora and Airglow, Academic Press, New York, 1961.

[28]Bates, D.R., in Physics of the Upper Atmosphere, ed. by J. A. Ratcliffe, Academic Press, New York, 1960.

[29]Fanale, F.P., Johnson, T.V., and Matson, D.L., "Io: A Surface Evaporite Deposit?", Science, Vol. 186, 1974, p. 922.

[30]Matson, D.L., Johnson, T.V., and Fanale, F.P., "Sodium D-Line Emission From Io: Sputtering and Resonant

Scattering Hypothesis", Astrophys. J. (Lett.), Vol. 192, 1974, p. L43.

[31]Gurnett, D. A., "Sheath Effects and Related Charged-Particle Acceleration by Jupiter's Satellite Io", Astrophys. J., Vol. 175, 1972, p. 525.

[32]McElroy, M.B., Y. L. Yung, and R. A. Brown, "Sodium Emission from Io: Implications", Astrophys. J. (Lett.), Vol. 187, 1974, p. L127.

[33]Sill, G. T., "Sources of Sodium in the Atmosphere of Io", presented at the 5th Annual Meeting of the Division of Planetary Sciences, AAS, Palo Alto, California, 1975.

[34]McDonough, T. R., and Brice, N. M., "New Kind of Ring Around Saturn", Nature, Vol. 212, 1973, p. 513.

[35]McDonough, T. R. and Brice, N. M., "A Saturnian Gas Ring and the Recycling of Titan's Atmosphere", Icarus, Vol. 20, 1973, p. 136.

[36]Lotz, W., "Electron-Impact Cross-Sections and Ionization Rate Coefficients for Atoms and Ions", Astrophys. J. (Suppl.) Vol. 14, 1967, p. 207.

[37]Intriligator, D.S. and Wolfe, J.H., "Initial Observations of Plasma Electrons from the Pioneer 10 Fly By of Jupiter", Geophys. Res. Lett., Vol. 1, 1974, p. 281.

[38]Carlson, R. W. and Judge, D. L., "Pioneer 10 Ultraviolet Photometer Observations at Jupiter Encounter", J. Geophys. Res., Vol. 79, 1974, p. 3623.

[39]McDonough, T. R., "A Theory of the Jovian Hydrogen Torus", to be published in Icarus, 1975.

[40]Judge, D. L. and Carlson, R. W., "Pioneer 10 Observations of the Ultraviolet Glow in the Vicinity of Jupiter", Science, Vol. 183, 1974, p. 317.

[41]Carlson, R. W. and Judge, D. L., "Pioneer 10 Ultraviolet Photometer Observations of the Jovian Hydrogen Torus: The Angular Distribution", to be published in Icarus, 1975.

[42]Fite, W. L., Stebbings, R. F., Hummer, D. G., and Brackmann, R. T., "Ionization and Charge Transfer in Proton-Hydrogen Atom Collisions", Phys. Rev., Vol. 119, 1960, p. 663.

[43]Kliore, A., Cain, D.L., Fjeldbo, G., Seidel, B.L., and Rasool, S.I., "Preliminary Results on the Atmospheres of Io and Jupiter from the Pioneer 10 S-Band Occultation Experiment", Science, Vol. 183, 1974, p. 323.

[44]McElroy, M.B. and Yung, Y.L., "The Atmosphere and Ionosphere of Io", Astrophys. J., Vol. 196, 1975, p. 227.

[45]Carlson, R.W., Bhattacharyya, J.C., Smith, B.A., Johnson, T.V., Hidayat, B., Smith, S.A., Taylor, G.E. O'Leary, B.T., and Brinkmann, R.T., "An Atmosphere on Ganymede from It's Occultation of SAO 186800 on 7 June 1972", Science, Vol. 182, 1973, p. 53.

[46]Brinkmann, R.T., "Mars: Has Nitrogen Escaped?", Science, Vol. 174, 1971, p. 944.

[47]Hodges, R.R.Jr., Hoffman, J.H., and Johnson, F.S., "The Lunar Atmosphere", Icarus, Vol. 21, 1974, p. 415.

[48]Dessler, A.J., in The Atmospheres of Mars and Venus, ed. by. J. C. Brandt and M. B. McElroy, Gordon and Breach, New York, 1968.

[49]Kuiper, G.P., "Titan: A Satellite with an Atmosphere", Astrophys. J., Vol. 100, 1944, p. 378.

[50]Trafton, L., "The Bulk Composition of Titan's Atmosphere", Astrophys. J., Vol. 175, 1972, p. 295.

[51]Trafton, L., "On the Possible Detection of H_2 in Titan's Atmosphere", Astrophys. J., Vol. 175, 1972, p. 285.

[52]Hunten, D.M., in The Atmosphere of Titan, ed. by D.M. Hunten, NASA SP-340, Scientific and Technical Information Service, Washington, D.C., 1974.

[53]Hunten, D.M., "The Atmosphere of Titan", Comments on Astrophys. Space Science, Vol. 4, 1972, p. 149.

[54]Trafton, L., "Titan: Unidentified Strong Absorptions in the Photometric Infrared", Icarus, Vol. 21, 1974, p. 175.

[55]Gillett, F.C., Forrest, W.J., and Merrill, K., "8-13 Micron Observations of Titan", Astrophys. J. (Lett.), Vol. 184, 19 , p. L93.

[56]Allen, D.A., and Murdock, T.L., "Infrared Photometry of Saturn, Titan, and the Rings", Icarus, Vol. 14, 1971, p. 1.

[57]Sagan, C., "The Greenhouse of Titan", Icarus, Vol. 18, 1973, p. 649.

[58]Pollack, J. B., "Greenhouse Models of the Atmosphere of Titan", Icarus, Vol. 19, 1973, p. 43.

[59]Danielson, R. E., Caldwell, J.J., and Larach, D.R., "An Inversion in the Atmosphere of Titan", Icarus, Vol. 20, 1973 p. 437.

[60]Strobel, D. F., "The Photochemistry of Hydrocarbons in the Atmosphere of Titan", Icarus, Vol. 21, 1974, p. 466.

[61]Briggs, F.H., "The Radio Brightness of Titan", Icarus, Vol. 22, 1974, p. 48.

[62]Low, F.J. and Ricke, G.H., "Infrared Photometry of Titan", Astrophys. J. (Lett.), Vol. 190, 1974, p. L143.

[63]Brinkmann, R. T., "Occultation by Jupiter", Nature, Vol. 230, 1971, p. 515.

[64]O'Leary, B. T., "Frequency of Occultation of Stars by Planets, Satellites, and Asteroids", Science, Vol. 175, 1972, p. 1108.

[65]Taberie, N., "Distribution of Hydrogen and it's Lyman-Alpha Intensity in the Atmosphere of Titan", Icarus, Vol. 23, 1974, p. 365.

MODELS OF THE JUPITER RADIATION BELT

W. N. Hess*

National Oceanic and Atmospheric Administration,
Boulder, Colorado

Abstract

A model of the Jupiter Radiation Belt is presented which
has electrons and protons diffusing in from the solar wind.
When they are in the region $1 < L < 5$, they lose energy by
synchrotron radiation. By matching the observed synchrotron
radiation radial distribution, a diffusion coefficient of
$D = 1.7 \times 10^{-9}(R/R_J)^{1.95} \sec^{-1}$ is determined. Particles diffusing into the Jovian magnetosphere at this rate should be significantly absorbed by the Galilean moons especially Io and
Europa. Calculations here say that off-equatorial particles
should be reduced several orders of magnitude as they diffuse
past these two moons. Particles which move in or very near
the magnetic equatorial plane would not be absorbed nearly as
much because they will be able to avoid hitting the moons most
of the time.

Introduction

Sloanaker[1] discovered decimetric radio waves radiation
coming from Jupiter. These radio waves have been vigorously
studied for the last 15 years and it is now well established
that they are due to synchrotron radiation from energetic electrons spiraling around the magnetic field lines in the Jovian
magnetosphere. There are several lines of evidence to demonstrate this.

Presented as Paper 73-565 at the AIAA/AGU Space Science
Conference: Exploration of the Outer Solar System, Denver,
Colo., July 10-12, 1973.
*Director, Environmental Research Laboratories.

1. Berge[2], using an interferometric technique, mapped
the spatial distribution of the radiation as shown in Fig. 1.
The radiation clearly comes from a region of space larger
than the planetary disc, indicating a radiation belt origin.

2. Decimetric radiation is polarized with $\vec{\varepsilon}$ vector
lying more or less parallel to the Jovian magnetic equator.

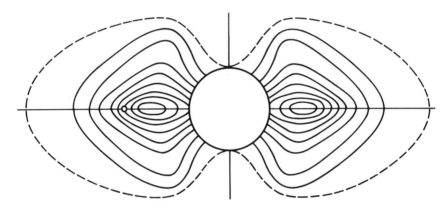

Fig. 1. The distribution of synchrotron radiation at 10.4
         centimeters wave length measured by a two-antenna
         interferometer by Berge[2]. The circle in the middle
         shows the disc of the planet Jupiter.

An early analysis by Chang and Davis[3] showed that electron fluxes of $J \sim 10^8$ elec/cm^2/sec of average energy $E \sim 10$
Mev would be required (assuming Jupiter had a magnetic field
of 10 gauss) to produce the observed radiation. They also
showed that the required electrons could not be diffused into
the Jovian magnetosphere rapidly enough to explain the observed effects by the process of variable solar wind pressure
producing magnetopause location fluctuations. Warwick[4], by
studying the decametric radiation from Jupiter, deduced that
it had a surface magnetic field of approximately 10 gauss.

Thorne[5] showed, by analyzing the polarization data and
the beaming of the radiation to the earth, that there must be
a very peculiar pitch angle distribution of the trapped electrons. His pitch angle distribution (where α_e is the angle
made by the electrons' resultant velocity vector to the local
magnetic field line) is

$$n(\alpha_e) = \cos^2\alpha_e + 2 \cos^{40}\alpha_e$$

The essential point here is that in order to have the $\vec{\epsilon}$ vector lie parallel to the equator most of the electrons must be of very nearly equatorial orbits. A $\cos^2\alpha_e$ distribution would clearly give the $\vec{\epsilon}$ vector lying parallel to the magnetic axis, not the equator.

From these studies we had a reasonably good idea about the electron radiation belt of Jupiter without having ever gone there. We have reasonable estimates of the planet's magnetic field, energetic electron flux, and average energy and pitch angle distribution in the inner part of the Jovian magnetosphere.

We had no idea about protons. Because of their heavy mass they are very inefficient radiators of synchrotron waves, so no remote detection of them is possible. Engineers at the Ames Research Center and the Jet Propulsion Laboratory responsible for the design of spacecraft which would fly close to Jupiter were worried by the possibility of very large energetic proton fluxes which might damage or destroy approaching spacecraft. A Jupiter workshop held at JPL[6] summarized opinions about the possible proton flux and produced the proton model shown in Fig. 2. The upper limit fluxes shown here are about enough to damage spacecraft unless substantial precautions are taken.

Because of this potential danger and also because of the extreme interest in the planet Jupiter caused by the launch of the satellite Pioneer 10, considerable effort has been devoted in the last two years to producing models of the electrons and the protons trapped in the Jovian magnetosphere. In this paper I will survey these models and show what seems to be the most likely picture of the trapped radiation for both protons and electrons.

Electrons

Because we know a good deal about the Jovian electron belt from the observed sychrotron radiation it is fairly easy to make a reasonable model of these particles. Almost all authors consider that the electrons originate in the solar wind and diffuse radially inward to the inner magnetosphere gaining energy as they go until they reach the region $1 < L < 5$ where they produce synchrotron radiation. The symbol, L, refers to the planet-centric distance (in Jovian radii) where a dipole-like field line intersects the magnetic equatorial plane. It is normally assumed that this inward radial diffusion conserves the first two adiabatic invariants: μ, the

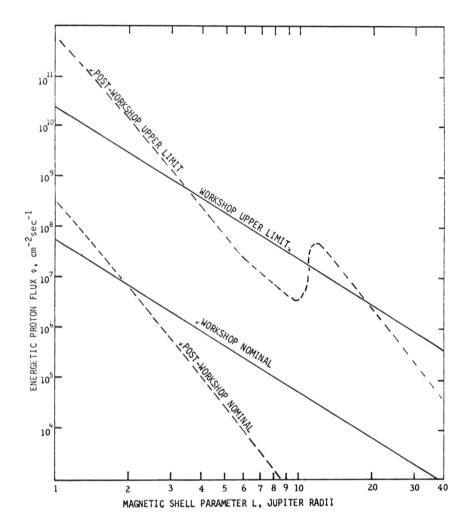

Fig. 2. Estimated proton fluxes as functions of distances
         from the surface of the planet in Jupiter's magnetic
         equatorial plane. These values were arrived at dur-
         ing a workshop on Jupiter held at the Jet Propulsion
         Laboratory in 1972.[6]

magnetic moment, and I, the integral invariant of the particle's motion. While this may seem like an arbitrary assumption, it is almost certainly true.

    Efficient radial diffusion, essentially, must conserve
μ and I. If there are diffusion-like processes that produce
arbitrarily directed changes in the velocity vector ΔV then

the center of gyration of a particle will be displaced by one gyroradius in about the same time that the direction of the velocity is changed by one radian. This means the particle will be precipitated before it diffuses radially much at all. The diffusion process must be highly constrained in order to avoid precipitation. The only reasonable process is one that conserves both μ and I.

Chang and Davis[3] produced the first good quantitative treatment of this problem which showed that magnetopause pumping could not produce the observed electron fluxes. It is now known that the diffusion-like process of fluctuating convecting electric fields also does not work. Brice and McDonough[7] have suggested that magnetic field line exchange (FLE) in the Jovian magnetosphere due to strong ionosphere winds will produce diffusion and may be the responsible process. This process qualitatively is uncertain. Jacques and Davis[8] have used this FLE process in constructing a diffusion model of the Jovian electrons. They find maximum fluxes of $2 \times 10^7/\text{cm}^2/\text{sec}$.

Stansberry and White[9] also use the FLE process and add, besides the loss due to synchrotron radiation, a second loss process which they need in order to match the observed electron distribution. They find a maximum electron flux of $J = 1.4 \times 10^9$ elec/cm^2-sec at L = 2.7 where the characteristic of energy is 7 Mev. They do not consider effects due to instabilities or lunar absorption.

Mead and Hess[10] showed what the characteristics of motion of trapped particles in the Jovian magnetosphere would be. Because the magnetic equatorial plane is inclined at 10° to the rotational equatorial plane, there is a special group of particles that should be able to diffusion radially inward more easily than the rest. As will be shown, absorption of particles by Jovian moons, Io and Europa, is very probably an important process. However, particles having near equatorial orbits will be able to avoid the moons, most of the time. Particles of equatorial pitch angle $\alpha_e = 90°$ would only be able to interact with the moon very near the node where the magnetic equatorial plane crosses the rotational equatorial plane. For this special class of near-equatorial particles having $\alpha_e > 69°$, lunar absorption is less important and therefore the population of these particles inside Io can be considerably larger than for particles with large bounce amplitude. This group of near-equatorial particles may explain the calculations of Thorne[5] who showed that there must be a very large population of very near-equatorial particles.

Coroniti[11] has also used FLE and developed a picture of
the electron flux diffusion inward from the Jovian magnetopause into the region of synchrotron loss. He uses a radial
diffusion coefficient of $D = 2 \times 10^{-5} L^3 R_J^2/$day in order to agree
with the radial distribution of synchrotron radiation. Coro-
niti[11] assumes in the outer portion of the radiation belt of
Jupiter, $7 < L < 20$, that the electron fluxes are controlled
by whistler wave precipitation, and that the electron flux is
limited to a stable trapped level where whistler wave growth
does not occur. Coroniti assumes, following Mead[12], that the
moons Io and Europa are efficient absorbers of electrons and
that only electrons with pitch angles of $90° \pm 10°$ will diffuse in past these satellites. In the inner zone, $1.2 < L
< 4$, a steady state radial diffusion equation with synchro-
tron energy loss is solved to give the relativistic proton
flux profile and mean energy.

BHNBL Model for Electrons

Birmingham et al.[13] have developed a quantitative model
of the Jovian electron belt without involving any one particle
diffusion *process*. They assume one source of particles popu-
lating the inner magnetosphere due to radial diffusion concerning μ and I. They assume two loss processes (a) due to syn-
chrontron radiation and (b) due to absorption at the surface
of Jupiter. They write a steady state transport equation for
this system of electrons as

$$\frac{\partial}{\partial R}\left[\frac{D}{R^2}\frac{\partial}{\partial R}(nR^2)\right] + \frac{\partial}{\partial \mu}\left(\frac{\partial \mu}{\partial t}n\right) = N\,\delta(R - R_1)\,\delta(\mu - \mu_1)\quad(1)$$

where n is the number of electrons between $\mu + d\mu$ contained
in a flux tube which crosses the equatorial plane at a distance R from the center of the planet. The first term in
Eq. 1 is the usual form of diffusion for this situation. The
second term represents energy loss by synchrotron radiation.

The right side of Eq. 1 represents a source of strength
N located in R_1 emitting electrons of magnetic moment μ_1. The
diffusion coefficient D in Eq. 1 is parameterized by

$$D = k\left(\frac{R}{R_J}\right)^m$$

where k and m are constants to be determined.

This diffusion equation is now solved to give $n(R,\mu)$ for
one particular set of the three parameters μ, k and m. This

solution is then compared with experimental data on the observed radial distribution of volume emissivity of synchrotron
radiation $I_0(R)$. This is done by deconvolving the observed
spatial distribution of synchrotron radiation from Jupiter as
measured by Berge[2] and shown in Fig. 1. BHNBL compare the observed values of $I_0(R)$ with their calculated values $I_c(R)$ obtained by

$$I_c(R) = \int_0^{\mu_1} n\,(R,\mu)\,d\mu \int_{-1}^{+1} E(R,\mu,\xi,\,\cos\alpha_e)\,d\,(\cos\alpha_e) \quad (2)$$

where E is the synchrotron power emission per electron per
frequency interval df centered at f from an electron of pitch
angle α_e and magnetic moment μ located at R.

The comparison of the observed $I_0(R)$ with the calculated
$I_c(R)$ allows the three parameters μ, k and m to be determined
quite well in a trial and error fashion. In this way, BHNBL
find that

$$\frac{D}{R_J^2} = 1.7 \times 10^{-9}\,(R)^{1.95}\,\text{sec}^{-1}$$

$$\mu_1 = 700\,\frac{\text{Mev}}{\text{gauss}} \quad (3)$$

This seems like a quite large magnetic moment but using the
relativistic forms

$$\mu = \frac{p^2}{2m\,B} = \frac{p^2 c^2}{2E_0 B} \quad (4)$$

$$p^2 c^2 + E_0^2 = (E + E_0)^2$$

we find that relativistically

$$E = 2E_0\mu B + E_0^2 - E_0 \stackrel{\sim}{\sim} 2E_0\mu B \quad (5)$$

and for μ = 770 Mev/gauss at R/R_J = 2.8 at periapsis for
Pioneer 10 the electron energy would be

$$E_e \stackrel{\sim}{\sim} 770\,\frac{4}{(2.8)^3} = 11.8\,\text{Mev}$$

which is not a very large energy. This diffusion coefficient
D clearly shows that the diffusion process that transports

electrons radially in the Jovian magnetosphere is not due to
disturbances at the magnetopause, which has a radial dependence of R^{10} (Chang and Davis[3]), or due to a fluctuating convective electric field, which has a radial dependence of R^6.
The dominant diffusion process may be field line exchange
driven by atmospheric-ionospheric winds as described by Brice
and McDonough[7], which goes as R^3, although this is not certain.
The rate of diffusion must be considerably slower than those
given by Brice and McDonough of $D = 6x10^{-8}L^3 sec^{-1}$ and by
Jacques and Davis[8] of $D = 5x10^{-8}L^2(L-1) sec^{-1}$.

The BHNBL diffusion coefficient has been determined empirically by fitting data and by using the simplest reasonable
physical model. It seems likely that it is roughly correct.
If other physical processes are present, such as the second
loss process of Stansberry and White[9], then the value of D
will change. But for the simplest model of the Jovian radiation belt, it is the best available value of D.

We will assume that the BHNBL value of $D = 1.7 \times 10^{-9}$
$R^{1.95} sec^{-1}$ is correct for both electrons and probably also
for protons in the rest of this paper.

Now we can consider the problem of absorption of diffusing electrons by the Galilean moons. Hess, Birmingham and
Mead[14] showed that the effect of the moons can be handled by
adding a loss term to the left side of Eq. 1 of

$$- \sum_{i=1}^{4} \frac{n}{\tau_i} S(R-R_i \pm a_i) \qquad (6)$$

This loss is due to absorption by the 4 moons (i = 1 to 4),
Amalthea (R = 2.55 R_J), Io (R = 5.95 R_J), Europa (R = 9.47 R_J),
and Ganymede (R - 15.1 R_J). The step function $S(R - R_i \pm a_i)$
is unity over the region, $R_i - a_i < R < R_i + a_i$ and zero else-
where. The average lifetimes from Mead and Hess[10] are

$$\tau \text{ Amalthea} = 2.43 \text{ d}$$
$$\tau \text{ Io} = 0.54 \text{ d}$$
$$\tau \text{ Europa} = 0.47 \text{ d}$$
$$\tau \text{ Ganymede} = 0.44 \text{ d}$$

Equation 1, with the addition of the lunar loss term Eq. 6,
has been solved numerically to give the data in Fig. 3. Electrons of μ_1 = 770 Mev/gauss are injected at a large distance
from Jupiter. It doesn't matter where this *outer* source is
placed, as long as it is far from the region of interest.
The electrons diffuse inwards and are partly absorbed by the

moons and then lose energy by synchrotron radiation in the region 1 < L < 6 to produce a radial distribution like curve A of Fig. 3. This distribution is of particles of constant μ, not energy. In this region close to the planet, an *inner* electron source effectively exists, produced by energy loss by the particles that have diffused outwards and behave like curve B of Fig. 3

Really we will not have a monoenergetic electron source, so lower energy electrons diffusing in from the *outer* source may have the same values of μ as electrons diffusing out from the inner source. The superposition of these two particle groups will make radial distributions intermediate between curves A and B of Fig. 3. The most distinctive feature of these curves will be a downward pointed cusp at the location of the moon.

The Problem of Protons

Protons in the Jovian magnetosphere cannot be studied at the earth because they produce almost no synchrotron radiation. So, before Pioneer 10 encountered Jupiter, there were only educated guesses (and some other guesses) about what trapped proton fluxes and energies were there. The energetic proton flux might be similar to the electron flux $J \sim 10^8$ $cm^{-2}sec^{-1}$, although there is no good reason why they should be similar. If the protons have the same magnetic moment as the electrons μ = 700 Mev/gauss, then they will have an average energy at $2R_J$ of $E = \mu B = [700] \, 4/(2)^3 = 350$ Mev.

There are several possible theoretical models, any one of which may describe the energetic proton flux:

(a) Diffusion Dominated by Magnetopause Pumping

If changes in the solar wind pressure, producing changes in the magnetopause, is the major particle diffusion process, as they seem to be for outer belt protons at the Earth[15], then there will probably be no protons close to the planet Jupiter. The reason for this is that the diffusion rate expected here is so slow that the Jovian moons Europa and Io will very completely absorb the radially diffusing protons (Mead and Hess[10]). There could be substantial fluxes of relatively low energy protons outside Europa, but they should be absorbed before . getting in past Io. It has been known for some time that magnetopause pumping does not work for the Jovian electrons[3]. The electrons cannot diffuse in fast enough this way to overcome the loss by synchrotron radiation.

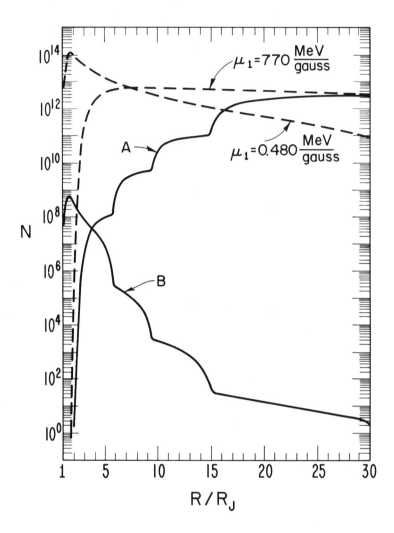

Fig. 3.   Relative electron fluxes at different distances from
          the center of the planet Jupiter. The dashed curves
          are what the fluxes would be if there were no moons
          of Jupiter. The solid curves are with the actual
          moons and show the substantial decreases due to the
          absorption by the moons. Curve A is for electrons
          of magnetic moment μ = 770 Mev/gauss, diffusing in-
          ward from a source at 30 R_J. Curve B is for elec-
          trons of magnetic moment μ = 0.48 Mev/gauss. These
          low energy electrons have been made by synchrotron
          radiation loss from the high energy source electrons.

(b) Io Source

It has been suggested[16] that there may be an energetic particle source related to a $\varepsilon = \vec{V} \times \vec{B}$ electric field extending across the moon Io. This ε field is due to the motion V of the moon through the magnetic field of Jupiter B. This process should produce electrons of E ∿ 300 Kev just inside Io and also protons of E ∿ 300 Kev just outside Io. These protons will have a magnetic moment

$$\mu \backsimeq \frac{E}{B} \sim \frac{0.3}{4/(6)^3} = 16 \text{ Mev/gauss} \tag{7}$$

so at $R/R_J = 2$ these protons would have E ∿ 8 Mev so they are not really very energetic. We will not consider them further.

(c) CRAND Protons

Protons from cosmic ray albedo neutron decay (CRAND) apparently do dominate in the terrestrial inner radiation zone[17]. Are they important at Jupiter? At first thought, they would appear to be unimportant because the Jovian magnetic field is considerably stronger than the Earth's field, so few cosmic rays can reach the planetary surface to make neutrons. However, the Jovian neutral atmosphere is quite thin so the lifetime of the trapped protons may be very long. The trapped proton flux from CRAND may not be very small at all because of this situation. The cosmic ray flux reaching the planet can be calculated roughly by using the verticle cut-off rigidities P_c for the planets. This cut-off gives the lowest momentum proton, which starting inwards vertically, can just reach the planetary surface at R

$$P_c = \frac{M \cos^4 \alpha_e}{4R^2} \tag{8}$$

where M is the planet's magnetic moment and α_e is the magnetic latitude. Using $M \propto B_0 R^3$ where B_0 is the surface equatorial magnetic field of 4 gauss we can write

$$\frac{P_c^{Jup}}{P_c^{Earth}} = \frac{P_c^{Jup}}{15 \frac{Bev}{c}} = \frac{B_0^J(R_J)}{B_0^e(R_e)} = \frac{4G}{.31G} \quad \frac{70000 \text{ km}}{6370 \text{ km}} \tag{9}$$

$$P_c^{Jup} = 2140 \frac{Bev}{c}$$

The integral proton energy spectrum in cosmic rays is

$$N(>E) = \frac{A}{(E+5.3)^{1.75}} \quad \text{with E in Bev} \quad (10)$$

Relativistically, $P_c c \simeq E$ so we can approximate the fraction F of the cosmic ray flux reaching the earth's surface that reaches the Jovian surface by

$$F = \frac{N_J(> P_c^{Jup})}{N_E(> P_c^{Earth})} = \frac{(P_c^{Earth} + 5.3)^{1.75}}{(P_c^{Jup} + 5.3)^{1.75}} = \frac{1}{3800} \quad (11)$$

So the Jovian CRAND source should be about 10^{-4} that of Earth. However, the loss rates of these Jupiter protons may be considerably smaller than for Earth.

The Jupiter atmosphere is practically non-existent at radiation belt altitudes. Ioannidis and Brice[18] calculate an equatorial cold plasma density of 0.2 electrons/cm^3 for $1 < L < 5$. This represents a reduction $\sim 10^4$ over the earth's atmospheric particle densities typically used to get the energetic inner zone proton flux from CRAND. This means that the energetic proton flux in the inner regions of the Jovian radiation belt due to CRAND might be expected to be the same order of magnitude as the inner zone proton flux for the earth having about the same energy spectrum. However, we have omitted radial diffusion here. When we consider that proton lifetimes would have to be thousands of years, clearly diffusions cannot be omitted. The time for a particle to diffuse a distance x is given approximately (Mead and Hess[10]) by

$$\tau = \frac{x^2}{4D} \quad (12)$$

If we ask how long it takes a particle to diffuse a distance $x = R_J$ and using $D = 1.7 \times 10^{-9} R^{1.95} R_J^2/\text{sec}$, we get

at L = 2 τ = 20 years
at L = 5 τ = 0.8 years

This means that almost all of the CRAND protons would have been absorbed in a time short compared to the slowing down time from atmospheric interactions. Because of this, the CRAND energetic proton flux at Jupiter will be so small that it can be ignored.

(d) BHNBL Model

Using the results of the radial diffusion model of Birmingham et al.[13] described earlier, the expected proton flux has been calculated (Hess, Birmingham, and Mead[14]). We don't know that the same processes apply to electrons, but it may be true and we have calculated the proton fluxes with this assumption.

We have used the electron transport equation from BHNBL and omitted the synchrotron loss term to give

$$\frac{\partial}{\partial R}\left[\frac{D}{R^2}\ \frac{\partial}{\partial R}\ (nR^2)\right] - \sum_{i-1}^{4}\ \frac{n}{\tau_i}\ S(R-R_i \pm a_i\ ) = N\ \delta(R-R_o)\ \delta(\mu-\mu_o) \tag{13}$$

This equation has been solved by a numerical finite difference method to give the results shown in Fig. 4. The protons do not change magnetic moment because there is no synchrotron loss. The data shown in Fig. 4 indicates large reductions in proton flux due to the moons. The reduction at Ganymede may not be observed because other processes may dominate for the outer magnetosphere, but the large reductions at Io and Europa and the small reduction at Amalthea should be real.

There is not the confusion present here that exists for electrons for external and internal sources. We have only an external source so all proton radial distribution curves for particles of one value of μ should look like Fig. 4.

Only the reduction expected due to the Jovian moons was determined by this study, not the absolute flux of protons.

(e) Other Models

Stansberry and White[9] have made predictions of the Jupiter proton fluxes based on a radial diffusion model of the type predicted by Brice and McDonough[7]. This model uses synchrotron radiation energy loss and another unexplained loss mechanism which seems to be required to make the electron model work. This model predicts a maximum proton flux of $1.8 \times 10^{10} \text{cm}^{-2} \text{sec}^{-1}$ at 1.3 R_J. At this point, the protons have a characteristic proton energy of 340 Mev. The characteristic proton energy changes with position here as

$$E \propto B \propto R^{-3}$$

Neither instabilities nor lunar absorption loss are considered here.

Coroniti et al.[19] developed an upper limit model of Jovian protons which allowed them to diffuse in radially with no

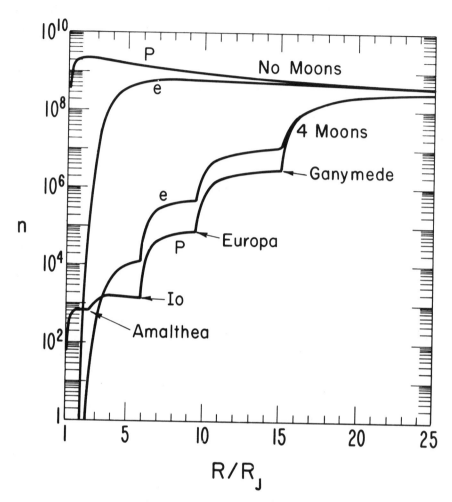

Fig. 4. Relative proton and electron fluxes as a function of
         distances from the planet Jupiter. The electron
         curves are the same as those shown in Fig. 3. The
         proton curves assume the same source strength and
         diffusion coefficient and omit synchrotron radiation
         to show that the absorption due to the moons of Jupi-
         ter is even larger for protons than for electrons.

losses with a diffusion coefficient $D = 2 \times 10^{-9} L^3 R_J{}^2$/sec up to
$L \sim 12$. Inside this radius the proton flux should be limited
by the ion cyclotron wave instability to be at the marginally
stable limit. In the range $1.5 < L < 5$ another instability
due to an electrostatic ion loss-cone wave further limits the
flux.

Several models of the Jovian electrons and protons have been presented here. Now let us summarize what we consider to be the most reasonable picture of Jupiter.

Pioneer 10 Flyby

On December 3, 1973, Pioneer 10 came within 2.8 R/R_J of the planet Jupiter. At the time this paper is being finished there are only fragmentary results from the experiments, but they do bear on this present paper so a short summary is in order.

(1) The magnetic field near the planet seems to be nearly dipolar with a surface field of $B_0 \sim 4$ gauss.

(2) At L = 3.2 the omnidirectional flux of electrons of E > 30 Mev is $J_e = 1.3 \times 10^7$ elec/cm^2-sec.

(3) At L = 2.8 the omnidirectional flux of protons of E > 30 Mev is $J_p = 4 \times 10^6$ protons/cm^2-sec.

(4) All four particle detectors show some satellite effects. Two of them seem to show good-sized dips at the location of I_0, while the other two seem to show good-sized dips at Ganymede and small effects at Io. It will take some time to sort out this data and look properly for the downward pointing cusps that cover a width of several R_J, as suggested in this paper. However, dips associated with the moons do seem to exist, as predicted here.

Conclusions

(1) On the basis of the simplest reasonable model of the Jupiter electron radiation belt, the radial diffusion coefficient for electrons has been determined empirically to be

$$D = 1.7 \times 10^{-9} R^{1.95} \qquad R_J^2/sec$$

(2) Using this value of D, the effect of electron absorption by the Galilean moons of Jupiter has been studied, and it is concluded that the effect should be large. Electrons having equatorial pitch angles $\alpha_e < 69°$ should interact strongly with the moons, especially Io and Europa. Reductions of the flux of inward diffusion electrons (outside source) of several orders of magnitude should occur. Low energy electrons produced close to Jupiter by synchrotron energy loss from higher energy electrons (inside source) will tend to diffuse outwards and suffer absorption by the moons, too.

well, one in wrong

(3) Near-equatorial electrons $\alpha_e < 69°$ will be absorbed less by the moons than off-equatorial electrons and can produce the strongly peaked pitch angle distribution that is required by the polarization data.

(4) If protons respond to the same diffusion process as the electrons, then they too will be quite strongly absorbed by the Galilean moons, and they too will have a near-equatorial peaked distribution. The protons should have no inner source as the electrons do. If protons do not respond to the electrons' diffusion process their fluxes probably will be even lower.

References

[1] Sloanaker, R. M., "Apparent Temperature of Jupiter at a Wavelength of 10 cm," Astronomical Journal, Vol. 64, 1959, p. 346.

[2] Berge, C. L., "An Interferometric Study of Jupiter's Decimeter Radio Emission," Astrophysics Journal, Vol. 146, 1966, p. 767.

[3] Chang, D. B. and Davis, L., Jr. "Synchrotron Radiation as the Source of Jupiter's Polarized Decimeter Radiation," Astrophysics Journal, Vol. 136, 1962, p. 567.

[4] Warwick, J. W., "Particles and Fields Near Jupiter," NASA Contractor Report CR-1685, Jet Propulsion Laboratory, Pasadena, California, 1970.

[5] Thorne, K. S., "Dependence of Jupiter's Decimeter Radiation on the Electron Distribution in its Van Allen Belts," Radio Science, Vol. 69-D, 1965, p. 1557.

[6] Divine, N., "Proceedings of the Jupiter Radiation Belt Workshop," Technical Memorandum 33-543, edited by A. J. Beck, Jet Propulsion Laboratory, Pasadena, California, 1972, p. 527.

[7] Brice, N. and McDonough, T. R., "Jupiter's Radiation Belts," Icarus, Vol. 18, 1973, p. 206.

[8] Jacques, S. A. and Davis, Leverett, Jr., "Diffusion Models for Jupiter's Radiation Belt," California Institute of Technology Preprint, November 10, 1972.

[9] Stansberry, K. G. and White, R. S., "Jupiter's Radiation Belts," Institute of Geophysics and Planetary Physics, University of California at Riverside, September 1973, 73-23 and Science, Vol. 182, 1973, p. 1020.

[10]Mead, G. D. and Hess, W. N., "Jupiter's Radiation Belt and the Sweeping Effect of its Satellites," Journal of Geophysical Research, Vol. 78, 1973, p. 2793.

[11]Coroniti, F. V., "Energetic Electrons in Jupiter's Magnetosphere," Astrophysics Journal, Vol. 191, No. 1, Part 1, July 1, 1974, p. 287 (abstract only).

[12]Mead, G. D., "Proceedings of Jupiter Radiation Belt Workshop," NASA-JPL Technical Memo 33-543, 1972, p. 271.

[13]Birmingham, T., Hess, W., Northrop, T., Baxter, R., and Lojko, M., "The Electron Diffusion Coefficient in Jupiter's Magnetosphere," Journal of Geophysical Research, Vol. 79, No. 1, Jan. 1, 1974, pp. 87-97.

[14]Hess, W. N., Birmington, T. J., and Mead, G. D., "Absorption of Trapped Particles by Jupiter's Moon," Journal of Geophysical Research, Vol. 79, July 1974, pp. 2877-2880.

[15]Nakada, M. P. and Mead, G. D., "Diffusion of Protons in the Outer Radiation Belt," Journal of Geophysical Research, Vol. 70, 1965, p. 4777.

[16]Shawhan, S. D., Gurnett, D. A., Hubbard, R. F., Joyce, Glenn, "Io Accelerated Electrons: Predictions for Pioneers 10 and 11," Science, December 28, 1973, p. 1348.

[17]Claflin, E. S. and White, R. S., "The Source of Inner Belt Protons," Journal of Geophysical Research, Vol. 78, No. 22, Aug. 1, 1973, pp. 4675-4678.

[18]Ioannidis, G. and Brice, N., "Plasma Densities in the Jovian Magnetosphere, Icarus, Vol. 14, 1971, p. 360.

[19]Coroniti, F. W., Kennel, C. F., and Thorne, R. M., "Stably Trapped Proton Fluxes in the Jovian Magnetosphere," Astrophysics Journal, Vol. 189, No. 2, Part 1, April 15, 1974, pp. 383-388.

Addendum (Added in Proof)

When this paper was written in 1973 there was no experimental data to use to see how good these models are. Then on
December 4, 1973, Pioneer 10 came within 1.84 Jupiter radii of
the planetary surface and on December 3, 1974, Pioneer 11 came
within 0.6 R_J of the surface.

From Pioneer 10 several detectors from three separate
experiments showed pronounced particle intensity dips at the
orbits of Io and Europa and perhaps small dips at the orbit of
Ganymede (see J. Geophys. Res. 79, No. 25, 1974). Detectors
on Pioneer 11, which reached inside the orbit of Amalthea,
showed intensity dips at the orbits of Amalthea and also other
unexplained peaks (Fillius, 1976[20]).

There is general agreement from the experiments on
Pioneer 10 that radial diffusion is a dominant process for
supplying energetic electrons and protons to the inner Jovian
magnetosphere. CRAND does not seem to be an important particle
source. There is evidence that Io is a source of energetic
particles (Fillius, 1976[20]) as suggested by Shawhan et al.
(1973[16]).

Pitch angle scattering seems to be a dominant loss
process (Fillius, 1976[20]) in the inner magnetosphere as well
as synchrotron radiation.

The Pioneer data have not yet been fully analyzed and
there is no quantitative comparison with data possible yet.
In general, Jupiter seems to be relatively earth-like in the
behavior of its inner radiation belt.

In general, it seems the data now available on the Jupiter radiation belt are in reasonably good agreement with the
radial diffusion and lunar absorption model presented in this
paper.

Reference Addendum (Added in Proof)

[20]Fillius, W., "The Trapped Radiation Belts of Jupiter," to be
published in Jupiter, the Great Planet, T. Gehrels, ed., University of Arizona Press, 1976.

PLASMA PHYSICS PHENOMENA IN THE
OUTER PLANET MAGNETOSPHERES

Frederick L. Scarf*

TRW Defense and Space Systems Group,
Redondo Beach, Calif.

Abstract

A source for Jupiter (or Saturn) radio-emitting particles
is the solar wind. The particle energies are raised to Mev as
plasma diffuses inward with magnetic moment conservation.
Wave-particle interaction phenomena play fundamental roles
here. Bow shock instabilities thermalize wind plasma, and
magnetospheric instabilities limit the trapped flux. Wave-
wave interactions probably account for the high intensity and
fine structure of decametric emissions, and instabilities can
produce shocks in front of supersonic satellites (e.g.,
Titan). Other phenomena are important at inner satellite orbits and along spacecraft trajectories; the energetic particles have $\kappa T \simeq$ Mev, and enormous plasma sheath electric
fields may develop.

Introduction

The earliest investigations conducted around the earth
from rockets and satellites were generally oriented toward the
disciplines of high energy particle physics, nuclear physics,
and cosmic rays. The earth's magnetic field configuration was

Presented as Paper 73-566 at AIAA/AGU Space Sciences Conference: Exploration of the Outer Solar System, Denver,
Colorado, July 10-12, 1973. The author thanks N. Brice, F. V.
Coroniti, R. W. Fredricks, E. W. Greenstadt, and J. Warwick for
helpful discussions about these topics. Analysis performed
under the auspices of the TRW Defense and Space Systems Group
Independent Research and Development Program. The material in
the first six sections was prepared in the Spring of 1973, well
before the Pioneer 10 flyby of Jupiter occurred. A brief
evaluation of these concepts, based on data obtained during
the Pioneer 10 encounter, is contained in the final section.
*Member of Professional Staff, Space Sciences Department.

analyzed, and the properties of the durably-trapped energetic particles were studied in detail. However, just a few years after the dawn of the space age, it became very clear that virtually all of the population of the Earth's energetic radiation belts was locally accelerated from the low energy solar wind, and that various collective processes were continuously responsible for precipitating particles from the trapped orbits. It is now known that the basic mechanisms that govern the dynamics of the Earth's magnetosphere involve plasma physics phenomena, and that most of the energetic trapped particles simply represent the high-energy tails of plasma distribution functions.

The central problems facing scientists concerned with the Earth's magnetosphere were summarized in recent reports of National Academy of Sciences study panels.[1,2] The reports emphasize the fact that the fundamental magnetospheric problems involve plasma convection and current systems, plasma instabilities, wave-particle interactions, and other collective phenomena that are directly associated with characteristics of non-equilibrium plasma distribution functions. This emphasis should not be surprising because the Earth's magnetosphere, in common with other astrophysical systems, is essentially a large scale plasma physics laboratory. Even in diverse fields such as solid state physics, effects of the solid state plasma are of major importance; here it is well known that the wave-like plasma oscillations of the electrons in the positive ion lattice play the dominant role in determining gross macroscopic characteristics of the crystalline state, such as superconductivity and ferromagnetism.

In recent years, the National Academy of Sciences conducted two additional studies specifically concerned with outer planet exploration, and in both instances the panelists strongly recommended that the prime scientific objectives of the exploration include study of magnetospheric and bow shock wave-particle interactions and spontaneous planetary emissions.[3,4] Despite these recommendations, based on information acquired in Earth orbit, the present approach to the study of outer planet magnetospheres has actually been oriented almost entirely in terms of high energy physics, to the exclusion of plasma physics. Experimental payloads have been put together chiefly to map planetary magnetic field configurations and trapped radiation profiles. High energy charged particles are important, of course, even in practical terms, and especially for Jupiter. That is, the radiometric emissions from Jupiter have been analyzed extensively to deduce the energetic electron characteristics, and it is widely recognized that these

trapped electrons and the associated protons present a potential flyby radiation hazard that can be extremely serious. In
fact, some have tried to justify the present high energy physics approach to study of the Jupiter magnetosphere on the
basis that the potential radiation hazard there is so severe
that the highest priority has to be given to analysis of the
energetic particles themselves, without regard to the mechanisms that produce them. However, this type of argument is
superficial and untenable. Although it is true that a very
great potential danger comes from the trapped proton fluxes
and that, since these particles do not produce any radiation
that can be detected from Earth, only _in situ_ observations
can conclusively shed light on the actual hazard, it is also
true that by the time local measurements reveal the extent of
the danger it is rather late to use the information for mission planning purposes.

Because of this experimental impasse, mission planners
did briefly turn to magnetospheric scientists with the recog-
nition that one can try to use basic plasma physics principles
to estimate the overall Jupiter radiation hazard by constructing fairly complete and self-consistent models of the origin
of trapped energetic particles in outer planet magnetospheres.
To this end the JPL Jupiter Radiation Belt Workshop was con-
vened two years ago, and a number of magnetospheric plasma
physicists were encouraged to develop comprehensive Jupiter
models from first principles of magnetospheric and plasma
physics. However, when the specific models presented in 1971
by Brice, Coroniti, Kennel, and Thorne[5] suggested that a wide
range of flyby trajectories could be safely negotiated, outer
planet exploration planning simply resumed its previous
course, with emphasis on studies of planetary atmospheres,
mapping of magnetospheric boundaries, and measurement of
planetary magnetic fields and energetic particle distributions, without acknowledging that plasma processes on which
speculative models are based would need experimental verifi-
cation too.

It is generally dangerous to rely completely on theoretical predictions because theoretical models, especially those
without a firm data base to work from, can and do change. In
fact, there have been some significant new developments in
this field suggesting that local plasma physics measurements
at the outer planets are now more appropriate than ever.
That is, the Jupiter Radiation Belt Workshop had very positive
aftereffects in that a number of space plasma physicists were
encouraged to continue research on Jupiter and Saturn magnetosphere models. In the past two years several varieties of

plasma physics models were developed further; the predictions
are conflicting and many of the new theories indicate that the
Jupiter radiation hazard is more severe than originally supposed. Moreover, it is now recognized that extremely large
plasma sheath electric fields may form around spacecraft subsystems, posing an additional complication on a flyby mission,
and presenting a need to understand the plasma environment.
This note contains: 1) a brief outline of the Radiation Belt
Workshop model for Jupiter and an associated generalization
for Saturn; 2) a summary of some newer concepts concerning
cold plasma distributions, stable trapping limits, and bow
shock interactions; 3) comments on recent ideas concerning the
plasma sheath around Io and Io-radio noise modulation; and
4) speculation about spacecraft charging problems near Jupiter.

The Radiation Belt Workshop Models
and Some Second Thoughts

The basic principles are illustrated by considering the
Earth's magnetosphere, which is sketched in Fig. 1. As noted
here, solar wind protons with streaming energy of about 750 ev
flow toward Earth, and the interplanetary magnetic field
strength is about 5 γ. The protons are heated and slowed down
by wave-interactions associated with plasma instabilities at
the bow shock; other wave-particle interactions in the magnetosheath and all along the magnetopause allow them to be
considered as magnetosphere injection sources with μ_p =
$E_p/B_{wind} \simeq 15$ Mev/gauss. Some particles presumably migrate
(via diffusion and convection) to low altitudes, conserving
μ_p, and hence E_p (L $\simeq$ 1-2) $\simeq$ 0.3 x 15 $\sim$ 5 Mev.

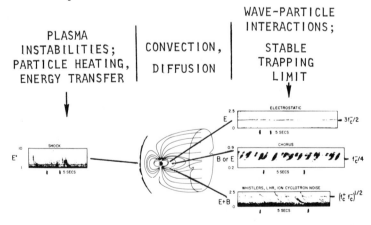

Figure 1   General outline of current models of
the Earth's magnetosphere.

It is frequently assumed that electrons are heated in much the same way, but this may be somewhat coincidental. The value of (κT_e) in the wind is about 10 ev, and fluid models give $T'(nose)/T(upstream) \simeq 15\text{-}30$. Thus, one might expect $\mu_e \simeq 3\text{-}6$ Mev/gauss. In fact, in the Earth's tail, T_e is clearly less than T_p.

The other important aspect of earth plasma physics theory concerns the concept that internal plasma instabilities will develop to limit the stably trapped flux. As particles drift in conserving μ_e, μ_p, then $T_\perp/T_\parallel$ should increase, triggering electromagnetic whistler mode noise (i.e., chorus or hiss) and ion cyclotron turbulence (see the right-hand side of Fig. 1). These waves will limit the flux of particles having energies higher than $(B^2/8\pi N)$. Other electrostatic instabilities (see Fig. 1) may also be very important.

The Jupiter-Saturn models recently developed are based on these general concepts, with a few new wrinkles. At Jupiter, B_0 is certainly very large (12 gauss rather than 0.3 gauss at Earth), and at Saturn use of $B_0 = 1$ gauss can be shown to give no problems.[6,7] At Jupiter and Saturn the magnetospheres are much larger because (NmV^2) is way down and $B^2/8\pi$ is way up. Figure 2 is a <u>scale</u> drawing of the Sun, along with similar portions of the Earth, Jupiter, and Saturn magnetospheres. The Jupiter magnetosphere is certainly the largest object in the solar system, and it is probable that the Saturn magnetosphere is the second largest. (It should be noted that the tiny dots represent the planets Jupiter and Saturn drawn to <u>scale</u>.)

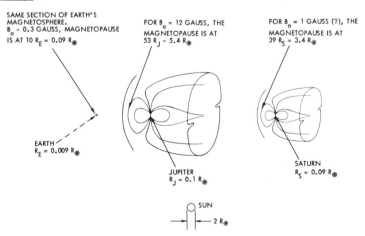

Figure 2   The Sun and planetary magnetospheres
(current models) to scale.

Since E_p is still about 750 ev at Jupiter and Saturn,
while the interplanetary field strength has drastically decreased, we find $\mu_p \simeq$ 100 and 200 Mev/gauss at Jupiter and
Saturn respectively. Thus if protons diffuse in to low L-
shells while conserving μ, they end up with extremely high
energies.

The Radiation Belt Workshop models assumed that: 1) the
high rotation rate (of Jupiter) would fling ionospheric photoelectrons out to give a specific cold plasma (Brice-Ioannidis
type) density distribution with $\beta \simeq 1$ at the magnetopause.
This means that a porous plasmasphere-type boundary would
develop, allowing magnetosheath particles to diffuse inward
readily; 2) some plasma instabilities in the shock-magneto-
sheath region would give $\mu_e \simeq \mu_p$; 3) the thermal anisotropies
associated with μ-conservation and inward diffusion would
produce certain stable trapping limitations.

The bottom panel in Figure 3 shows one prediction of the
Workshop. The upper limit electron flux and the Brice-
Ioannidis density distribution are both represented here. It
can be seen that very high electron energies and flux values
are predicted here, and the companion proton prediction does
indicate a serious hazard. However, somewhat different parameters yielded a less severe "nominal" model, and the Hess-
Mead[5] concept of satellite sweeping could be used to predict
even lower fluxes.

In fact, most developments since the Workshop appear to
lead to predictions of higher flux and fluences. The central
panel of Fig. 3 shows an energetic electron flux profile recently computed by Coroniti, Kennel, and Thorne.[8] Here the
sweeping effects of the satellites are taken into account,
and improved diffusion calculations are used. However, the
basic concepts of the Workshop models are retained (e.g., the
porous boundary, near-equality of μ_e, μ_p, and the Brice-
Ioannidis 5 ev photoelectron density distribution) and a number of internal electrostatic and electromagnetic plasma
instabilities are invoked to explain why the stably trapped
flux is so low.

The top panel in Figure 3 shows how the energetic electron flux would jump if no internal plasma instabilities
limited the trapped flux. In fact, several models suggest
that this situation could occur if the satellites are non-
conducting. For instance, Brice and Mc Donough[9] recently re-
calculated the cold densities taking into account the
sweeping effects of the Jovian satellites and their roles in

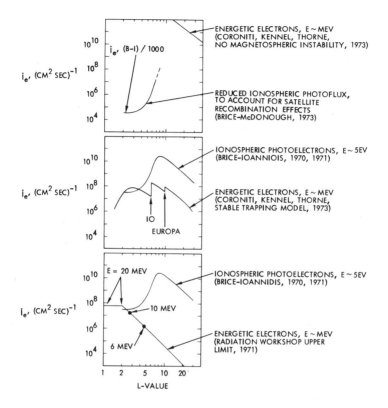

Figure 3   Hot and cold electron flux profiles
for various models.

enhancing the rate of recombination. Since the cold plasma is
essentially confined to the spin plane by the high centrifugal
forces, they found that the satellites can be very effective
in removing cold plasma, and the Brice-Ioannidis function
could then be an overestimate by orders of magnitude. The
curve sketched in the top of Fig. 3 is an illustrative one, in
which the Brice-Ioannidis function is simply reduced by an ad
hoc factor of 10^3. If the cold plasma density is actually so
low, several very important changes can be contemplated: a)
since $B^2/8\pi N$) is now much higher the cyclotron resonance instability now causes stable trapping limitations only for the
very energetic particles in the tail of the distribution, and
it can be effectively ignored; b) since $\beta = 8\pi N(cold)/B^2$ is
greatly reduced, it is not clear that the magnetopause
boundary will be as "porous" as assumed at the Radiation Belt
Workshop. Thus, we may end up with lower fluxes of higher
energy particles; c) if $j_e(cold) \ll j_e(hot)$ as sketched at the

top of Fig. 3, then ·the effective plasma temperature is extremely high. The implications of this possibility will be
taken up again in a later section.

The uncertainties in these models are actually enormous,
because of the interdependence of so many complex phenomena.
If the satellites of Jupiter are sufficiently conducting,
rather than non-conducting, one might expect the field lines
to be excluded, so that vastly reduced sweeping effects would
occur, and in this case it might be reasonable to utilize the
higher density Brice-Ioannidis type curves. However, even
with a model such as the one presented in the central panel of
Fig. 3, major questions have recently been posed. For instance, Michel and Sturrock[10] proposed that the rapidly rotating high-β plasma in the outer magnetosphere will cause the
magnetosphere to open up to form a planetary wind. This wind
would collide with the solar wind to form a fundamentally different kind of planetary interaction, controlled by a two-
stream plasma instability at the outer boundary. In another
area, even if one retains the conventional earth-type magnetospheric configuration shown in Fig. 2 for Jupiter or Saturn,
it is not at all clear that the electrons and protons are tied
to each other by $\mu_e \approx \mu_p$, as implicitly assumed at the Radiation Belt Workshop. At Jupiter and Saturn the solar wind
electrons may have thermal energies near 1 ev or less, and if
the electron temperature jump across the magnetosheath is governed by fluid concepts, rather than by plasma instabilities,
the maximum value for μ_e might be as low as (1.5-3) Mev/gauss,[7]
while μ_p could be near (100-200) Mev/gauss. Thus, the proton
hazards could conceivably be considerably worse, relative to
the electrons, than suggested two years ago. On the other
hand, Birmingham et al.[11] recently deduced a μ_e-value of 500
Mev/gauss, suggesting that extremely strong and unusual
electron-proton wave-particle interactions do take place upstream from the Jupiter magnetopause. This result might also
imply that the subsolar magnetopause is <u>not</u> the origin of the
energetic electrons. Coroniti et al.[8] considered models with
an electron source in the Jupiter tail, and these models give
$\mu_e \approx$ 500-1000 Mev/gauss. Thus, there is presently an uncertainty of a factor greater than 100 in assessing the importance of wave-particle interactions in heating electrons upstream from the magnetopause. If the wind is the source, the
one point that seems to be certain is that some remarkable
combination of local acceleration processes acts to raise electron energies from 0.5 to 1 ev (in the solar wind at 5 AU)
to about 20,000,000 ev in the inner belt of Jupiter, despite
the fact that these electrons are continuously losing energy
by radiating electromagnetic waves.

The basic questions that arise involve the roles of plasma instabilities in providing particle heating, energy exchange, radial diffusion, pitch angle scattering, and stable trapping limits. In all discussions of these phenomena, the cold plasma density profile plays a crucial role. As noted, inclusion of satellite sweeping and recombination effects can drastically reduce the nominal cold density distribution.

Satellite-Magnetosphere Interactions

It is well known that almost all of the natural satellites of Jupiter and Saturn move slowly in comparison with the local corotation speed, so that if the magnetospheres do corotate, the relative motions are retrograde with fairly high orbital speeds. Moreover, if the cold plasma has $\kappa T_e \simeq \kappa T_p \simeq 5$ ev, then most of the satellites move supersonically with respect to the protons and wake cavities (similar to the lunar cavity) should form. It is of interest to note that Titan moves at a relative speed of about 200 km/sec with respect to the plasma. It is likely, therefore, that this satellite, which possesses a detectable atmosphere, also has a bow shock and some sort of ionosphere, as at Venus. However, the very low plasma and atmospheric densities anticipated imply that the collisional mean free paths are huge, so that the Saturn magnetosphere-Titan atmosphere interaction would have to be governed by collective effects involving plasma waves.

The outer planet satellite that has received the most attention is Io because of its role in modulating the very intense decametric radiation from Jupiter. It is widely accepted that collective plasma interactions play a basic role in the generation of the decameter bursts; the source is very small and the radiation is much too strong (equivalent brightness temperature $\simeq 10^{14}$ to 10^{18}°K!) to be produced by any conceivable incoherent radiation process. Moreover, the decametric bursts are observed to have a millisecond fine structure, indicating that local plasma waves somehow interact at the source to produce coherence effects that can account for the high intensities. Since the position of the satellite Io influences the decametric radio emissions, recent theories of the Io-modulation effect therefore involve analyses of the Io-induced plasma instabilities.

Several of these theories are based on the observation that huge $(\underset{\sim}{V} x \underset{\sim}{B})$ electric fields will develop across the satellite as the Jupiter magnetosphere corotates past Io. For $B_0 \simeq$ 10 gauss, the potential difference across Io will be near 700,000 volts; Piddington and Drake[12] and Goldreich and

Lynden-Bell[13] assumed that this potential is transmitted unat-
tenuated along the magnetic field lines connecting Io to the
ionosphere, providing an auroral-type arc at the foot of the
field line. This type of explanation has been criticized on
general plasma physics grounds because the impressed electric
field is larger than the so-called runaway field.[14] In this
case current-driven plasma instabilities should develop, wave-
particle scattering should give rise to an enhanced or turbulent plasma resistivity, and the voltage drop across Io should
not be impressed across the ionosphere without attenuation.

Gurnett[15] recently suggested that the plasma sheath
around Io is a space-charge region where most of the voltage
drop occurs, and a sketch of his model is contained in Fig. 4.
The novel concept introduced by Gurnett is the idea that
photoelectrons emitted from Io will be locally accelerated in
this plasma sheath, attaining some fraction of the 700 kev
potential difference across the satellite. Presumably these
accelerated photoelectrons flow parallel to B, intersecting
the ionosphere at the Io field line, and thus producing modulation of decametric bursts.

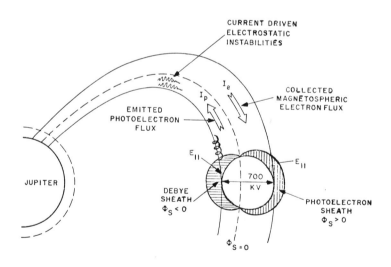

Figure 4   Schematic representation of Gurnett's
            model for Io modulation associated
            with field-aligned current systems
            and local acceleration of photoelec-
            trons in Io's plasma sheath.

   Recent local observations of wave-particle and wave-wave
interaction phenomena in the Earth's auroral region provide
some insight into the kind of phenomena that may occur along
the field lines connecting Io to the Jupiter ionosphere.
Figure 5, taken from a recent paper by Fredricks et al.[16] shows
OGO-5 wave and field observations in the region of the Earth's
dayside polar cusp. This cusp connects directly to the auroral
oval, and it contains relatively energetic magnetosheath plasma. Strong field-aligned currents do flow along the cusp
boundaries, and large amplitude plasma waves are radiated by
two-stream instabilities. Apparently the plasma waves do produce turbulent resistivity so that the field lines are not
equipotentials, and voltage drops along the auroral field lines
then cause local acceleration of auroral particles.

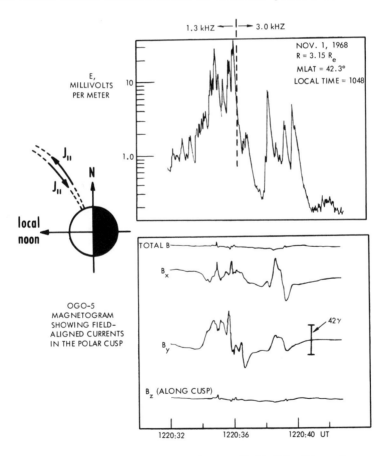

   Figure 5   OGO-5 observations of field-aligned cur-
              rent systems and associated plasma waves
              in the Earth's dayside polar cusp.

The earth's auroral region is also a source of high frequency electromagnetic radiation. This auroral hiss is generally attributed to some Cerenkov radiation process, but just
as with the Jupiter decametric bursts, the observed intensity
is now known to be too high to be explained in terms of incoherent radiation from the observed particles.[17] Recently,
Scarf et al.[18] suggested that the plasma waves associated with
the current-driven instabilities interact with the Cerenkov
radiation to produce coherent effects that account for the high
intensities. It does seem likely that the plasma physics processes in the Io flux tube are similar to those occurring in
the Earth's auroral region. However, it should be noted that
very different interpretations have also been proposed. For
instance, Wu[19] recently argued that sharp density gradients in
the energetic protons (caused by the sweeping effect of Io)
can generate drift-type instabilities that may be relevant.
The only certain conclusion is that the Io modulation problem
will not be solved unless the conditions near Io and its flux
tube are studied from the point of view of plasma physics.

Energetic Particles as an Ultra-High Temperature Plasma

Astrophysicists customarily discuss properties of extremely hot plasmas, including systems that have relativistic
thermal characteristics. However, until the last one or two
years those concepts were regarded as more or less theoretical
notions. Even if plasma temperatures of tens, hundreds, or
thousands of kilovolts did develop in nature, this was supposed
to happen in distant galaxies, and certainly not in our solar
system, or in the Earth's magnetosphere surrounding man-made
spacecraft payloads.

Of course, the theoretical ideas discussed earlier (of
convection and inward diffusion with conservation of μ) could
have been used several years ago to predict extremely high
plasma temperatures at a few Earth radii. For instance, if
electrons with $\mu_e \simeq$ 5-10 Mev/gauss convect or diffuse in to
L = 6, they should arrive with average energies of the order
of 5-10 kev. Such energetic electrons were indeed detected by
instruments on OGO-1, OGO-3, but it was always assumed that
these particles represented the high energy tail of the total
electron distribution, with a much denser but unmeasured cold
population being supplied from the ionosphere. Even when
workers who analyzed micropulsations and local ion data from
the OGO-5 spectrometer reported exceptionally low cold plasma
densities beyond the plasmapause, few observers interpreted
this to mean that the average thermal energy was in the kilovolt range.

Direct and conclusive evidence of plasma temperature values in the kilovolt range was first supplied by ATS-5 plasma probe experimenters. As before, they reported mean electron and proton energies in the kilovolt range, but in addition, De Forest[20] was able to show that the magnitude of the spacecraft potential (relative to the plasma) was in the kilovolt range for surfaces that were not exposed to sunlight. This conclusively proves that the effective plasma temperature is of the same order, because if secondary emission, ram effects, and backscattering are neglected, unilluminated surfaces in a plasma should develop a negative potential with

$$\left| e\phi(\text{dark}) \right| \simeq \kappa T_e \log(j_e/j_p)$$
$$\simeq \kappa T_e \ln \left[(T_e/T_p)^{1/2} (m_p/m_e) \right]$$
$$\simeq 4(\kappa T_e) . \tag{1}$$

De Forest primarily analyzed data from a plasma probe mounted just behind a conducting "belly-band" on ATS-5, and in this case the conducting surface always had a well-defined potential relative to the plasma, and that potential was determined by the instantaneous current balance at all points. That is, in general, one side of this conducting band was exposed to sunlight, and the positive current was primarily associated with emission of photoelectrons, so that Equ. (1) could not be applied. When ATS-5 was in sunlight, the potential of the conductor was approximately

$$\left| e\phi(\text{sun}) \right| \simeq \kappa T_e \log \left[j_e/j(\text{photo}) \right] \tag{2}$$

with $j(\text{photo}) \simeq 8.2 \times 10^{-10}$ amp/cm (again secondary emission effects are neglected here). The striking changes in potential relative to the plasma occurred when the entire spacecraft entered eclipse, and suddenly the charging phenomena would be described by Eq. (1) rather than by Eq. (2).

The top part of Fig. 6 shows some actual data presented by De Forest to illustrate this point, but the labels in Fig. 6 represent an extrapolation that will be explained shortly. The electron and proton distribution functions associated with open circles (labeled "sunlit side") were actually measured at all angles (near the conducting surface) when the entire spacecraft was in sunlight, and $\left| e\phi \right|$ was determined to be quite low. As the spacecraft entered eclipse the apparent spectra shifted drastically (see the X-marks). Electrons appeared to have lower energies, and De Forest deduced that this shift was caused by a jump in potential of the conducting band to -4200 volts.

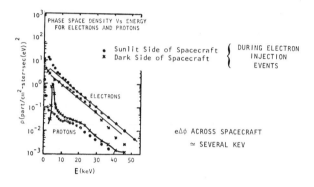

EXTRAPOLATION OF ATS-5 CHARGING ANALYSIS
FOR NON-CONDUCTING SPACECRAFT IN A PLASMA WITH

$j_e(\text{HOT}) \gg j_e(\text{COLD}),\quad \kappa T_e(\text{HOT}) \simeq 10\text{-}20 \text{ keV}$

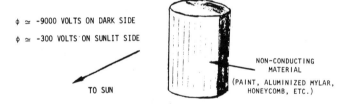

SPACECRAFT NOT AN EQUIPOTENTIAL.
POSSIBLE CONDITIONS:

$\phi \simeq -9000$ VOLTS ON DARK SIDE

$\phi \simeq -300$ VOLTS ON SUNLIT SIDE

NON-CONDUCTING
MATERIAL
(PAINT, ALUMINIZED MYLAR,
HONEYCOMB, ETC.)

TO SUN

Figure 6   Top: Observations of spacecraft charging to
kilovolt levels in synchronous Earth orbit
orbit during substorms (see text for complete
explanation). Bottom: Illustration that
electric fields of several kilovolts/meter
develop across non-conducting spacecraft subsystems.

The labels at the top of Fig. 6 represent an extrapolation of these measurements to nonconducting parts of the
same spacecraft. Illuminated surfaces will charge only moder-
ately, while shadowed surfaces will charge to very high negative potentials, as indicated in the bottom part of the figure.
Thus electric fields with E ≃ hundreds to thousands of volts/
meter will develop whenever $j_e(\text{hot}) \gg j_e(\text{cold})$ so that κT_e is
effectively in the range of one or more kilovolts.

De Forest did also analyze the response from another plasma probe mounted away from the conducting band, and he presented direct evidence that nonconducting surfaces produced local charge variations with associated electric fields of several hundred volts per meter, when κT_e was several kilovolts. More recently, Fredricks and Scarf[21] used engineering data from other synchronous spacecraft with nonconducting outer surfaces to infer the presence of local sheath electric fields exceeding one kilovolt/meter during noneclipse substorm injection events. Flight data and laboratory simulations indicated that portions of surfaces of a spacecraft not only charge to many kilovolts negative, but that they also suffer discharges (arcs or coronas). The large amplitude electromagnetic pulses with high frequency spectra irradiate cabling, and cause anomalous changes of state of electronics subsystems, degradation of aluminized mylar super insulating material, degradation of optical systems, etc. The association of spacecraft charging with spacecraft problems is not really as new as it might appear from this discussion. In 1959, Warwick[22] proposed that fluctuations in Sputnik 1 spin decay could be associated with high and asymmetric sheath fields as the satellite traversed the auroral zone. To summarize, in energetic plasma regions around the earth, differential charging of nonconducting spacecraft to kilovolt/meter levels has already been shown to give rise to a number of serious spacecraft and subsystem problems. It takes little imagination to anticipate what might happen at the outer planets where the corresponding plasma energization processes (i.e., inward convection or diffusion with μ-conservation) lead to prediction of $\kappa T_e \simeq$ hundreds to thousands of kev, rather than the modest 10-20 kev encountered in synchronous earth orbit during substorms.

Differential Charging and the Io
Modulation Effect

The nominal Radiation Belt Workshop Model[5] predicts that electrons diffusing in from the solar wind will have a characteristic energy (E_0) near 900 kev when they reach the orbit of Io. This E_0-value is only about a factor of 50 greater than the electron thermal energy actually measured in synchronous Earth orbit during substorms, and it seems appropriate to regard this energy as the equivalent local electron temperature for the plasma of solar wind origin. Presumably Io is also immersed in some cool plasma formed by ionospheric photoelectrons, and on its sunlit side, Io will emit (and reabsorb) additional low energy photoelectrons. Differential charging of this satellite of Jupiter will certainly occur if the outer surface is nonconducting, as the Earth's Moon is.

Figure 7 contains a simplified drawing of Io, indicating that the satellite outer surface should be subdivided into four major sections, in order to estimate crudely the surface potentials with respect to the plasma. On the sunlit faces the positive current will probably be associated with photoemission, while in the dark hemisphere j_p will come from collection of ambient protons (again in this simplified model we ignore secondaries, backscattering, and ram effects). There are also asymmetries in collection of negative currents. In the upstream region the satellite will be able to collect cool electrons (i.e., $\kappa T_e \sim 5$ ev photoelectrons from the ionosphere as in the Brice-Ioannidis model), very hot electrons ($\kappa T_e \sim E_o$, as in the Radiation Belt Workshop model), and some returning photoelectrons from Io itself. However, if a wake cavity forms, then in this wake region the cool Brice-Ioannidis electrons should essentially be absent, and κT_e(effective) could conceivably be near $E_o \simeq 900$ kev.

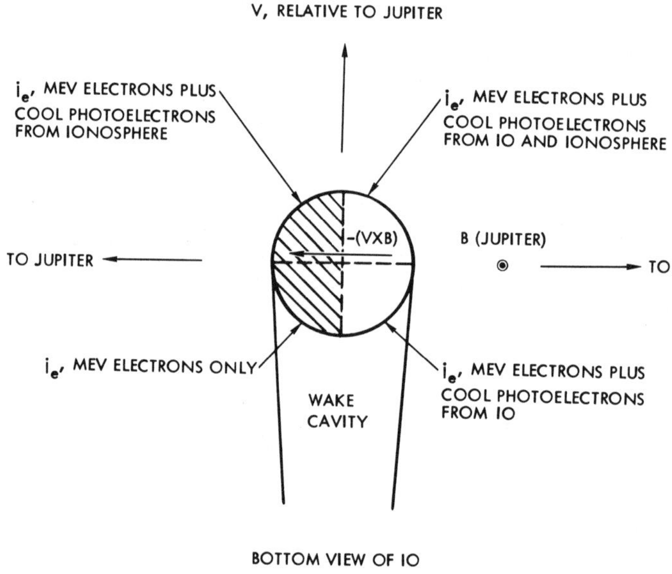

Figure 7   Simplified model for differential charging
           of Io. In at least one quadrant there are
           no cool (~ 5 ev) electrons, and κT_e(effective) $\simeq$
           1 Mev, so that the nonconducting surface
           acquires a huge charge.

In Fig. 7 we do indicate a possible difference between the wake-shadow and wake-sunlit portions of the Io surface, primarily to caution the reader that many details are still lacking here. However, the main point, which is illustrated in Fig. 7 is the following: even if j_e(cool) > j_e(hot) in the upstream region, the wake cavity should have a greatly reduced flux of cool plasma, κT_e(effective) in the wake may approach one Mev, and the differential charging can then give rise to an electric field across Io with a total potential difference near a million volts. Moreover, this sheath electric field would have a finite component of $\underset{\sim}{E}$ parallel to $\underset{\sim}{V}$(orbital), and so power would be fed directly into the surrounding plasma as the magnetosphere corotated past the satellite.

Figure 8 contains one generalization of these ideas to show how the effective electron temperature in the Jupiter spin plane would vary with L-value for a very specific model of the plasma environment. It is assumed here that the energetic electron flux is correctly given by the recent Coroniti, Kennel, Thorne calculation, and that the energy variation is described by the nominal Radiation Belt Workshop model. It is also assumed that in all regions except the satellite wakes, the cool photoelectrons ($\kappa T_e \sim 5$ ev) are correctly given by the Brice-Ioannidis model. Finally, it is assumed that in the satellite wake regions the cool plasma density is reduced by a factor of 1000 (Explorer 35 shows solar wind density depletions in the lunar cavity by at least factors of several hundred).

As shown in Fig. 8, with this model the effective plasma temperature would be very low except in regions where j_e(hot) exceeds j_e(cool). The effective electron temperature would rise drastically to between 300 kev and 6 Mev in the satellite wake regions (only Io and Europa are shown) and in the inner belt (L $\simeq$ 2-3.5, for this model). It should be noted, however, that this prediction of fairly limited spatial regions with very high temperatures is a rather optimistic one. For instance, if one uses the new reduced cold plasma density distributions of Brice-McDonough or Axford, or the higher μ_e = 500 Mev/gauss injection value recently deduced by Birmingham et al., the region with $\kappa T_e \simeq$ hundreds to thousands of kilovolts will spread over the entire inner magnetosphere of Jupiter. These considerations suggest the very real possibility that Io-induced modulation effects may be driven by differential charging of the satellite, rather than by $\underset{\sim}{V} \times \underset{\sim}{B}$ fields.

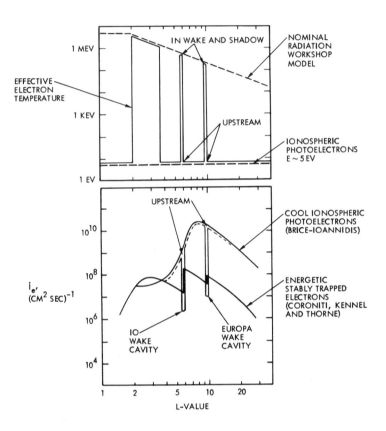

Figure 8   Idealized (and optimistic) model indicating
            where very high effective temperatures will
            be encountered during a Jupiter flyby in the
            spin equatorial plane. Away from the
            equator, j_e(hot) $\gg$ j_e(cool), and κT_e(effective)
            is huge everywhere.

Differential Charging of Flyby Spacecraft and Subsystems

The 700 kev potential difference that develops across Io
because of its motion through the Jupiter magnetosphere is
given by $\Delta\phi = (V \times B) \cdot \ell$, where ℓ is the diameter of the satellite. A small spacecraft in the same orbit would develop only
a modest potential difference because of this $(V \times B)$ electric
field effect. However, differential charging presents a

problem that is independent of the size of the object. If two
subsystems on a spacecraft are not electrically connected, they
will acquire a potential difference

$$e\Delta\dot\phi = e(\phi_2 - \phi_1) = \Delta\left[\kappa T_e \, \ell n \, j_e/j_p\right] \tag{3}$$

where the Δ on the right-hand side of Eq. (3) depends on the
difference in the external plasma and photoillumination characteristics.

Even if the relatively high cold density of the Brice-
Ioannidis model is correct, it is easy to see that there are
several regions where j_e(hot) $\gg$ j_e(cool), and κT_e(effective)
is E_0(energetic), so that

$$e\left|\phi(\text{illuminated}) - \phi(\text{shade})\right| \simeq$$

$$E_0\left[\ell n \, j_e/j_p(\text{photo}) - \ell n \, j_e/j_p(\text{plasma})\right] \tag{4}$$

In this case, for $E_0 \simeq$ hundreds to thousands of kev, very large
potential gradients will be impressed across the spacecraft,
presenting a possible hazard, even if the difference between
the two logarithmic factors is small. Some regions where this
complication is almost sure to be important are:

a) The inner belt of Jupiter, as shown in Fig. 8, for
$L \lesssim 3.5 \; \kappa T_e$ should be very high with almost any model.

b) In the satellite wake regions: see Fig. 8.

c) At a Saturn encounter, assuming B_S is of the order of
one gauss:[7] the reason for this is that in any model of the
cold density distribution, the cold electrons are essentially
confined to the spin equatorial plane, more or less as the
rings are. However, Saturn's spin axis is 27° from the ecliptic, and a flyby spacecraft will approach in the ecliptic
plane. Thus, the spacecraft will only intersect the spin
equator (and the region containing high fluxes of cool electrons) at isolated points in the trajectory - everywhere else
κT_e will be extremely high.

d) On a high inclination flyby of Jupiter: if $B_J \simeq 10$
gauss and if $\mu_e = 500$ Mev/gauss,[11] then even at $L = 20$, the
energetic particles will have $\kappa T_e \simeq 600$ ev, while the cold particle flux will be negligible beyond one or two Jupiter radii
from the spin equator.

Summary

As noted in the Introduction, although the very earliest explorations of the Earth's magnetosphere were carried out from a high energy physics point of view, the focus of attention soon shifted to the plasma physics discipline. After suitable instrumentation was developed, virtually all magnetospheric spacecraft payloads included some equipment to measure plasma distribution functions and plasma wave spectra. For instance, the OGO 1, 2, 3, 4, 5, 6; IMP 6, 7; ISIS 1, 2; Injun 5; S3-A; HEOS A-2, UK-4; Intercosmos 5; and Prognoz 1, 2 spacecraft all carried plasma wave and thermal particle measuring instruments into Earth orbit. All pending magnetospheric science missions (IMP-J, Hawkeye, Mother-Daughter-Heliocentric, GEOS) also have very comprehensive plasma physics payloads.

This historical lesson has been misinterpreted or simply ignored in establishing priorities for outer planet exploration. It is ironic that the high energy physics approach is given such high priority because, in fact, from radio observations we already know much more about the inner belt of Jupiter than we knew about the Earth's trapped population when the payloads for the earliest Explorers and Pioneers were being put together. On the other hand, we know virtually nothing about the origin of these energetic particles. In a very real sense the Radiation Belt Workshop models have fallen apart in the past two years. New theories predict much lower cold plasma densities and raise basic questions about stable trapping limits. New interpretations of the radio observations imply such huge injection values for μ_e that the numbers seem to rule out injection from the subsolar magnetopause, unless tremendously efficient wave-particle acceleration phenomena occur within the magnetosphere.

Perhaps the emphasis on high energy physics came about because of concerns with radiation damage, but even years ago it should have been recognized that other possible spacecraft hazards would be encountered. It is a simple matter to calculate from ground-based Jupiter radio observations that a flyby spacecraft will be immersed in RF wave fields having amplitudes of several volts/meter. Moreover, although there is no direct knowledge of the wave amplitudes at frequencies much below 10 MHz, we know that most wave spectra in nature have much higher amplitudes at the lower frequencies.

In 1973, it is also clear that certain plasma physics phenomena associated with the spacecraft sheath can present as much danger for a Jupiter flyby mission as the high radiation levels do. Spacecraft, subsystems, and scientific instruments should be developed with all those potential hazards in mind, and balanced payloads should also be designed to provide complete and unambiguous information about the spacecraft environment and scientific instrument operation, as well as data on the fundamental natural processes that occur in the outer planet magnetospheres.

References

[1] Priorities for Space Research, 1971-1980, National Academy of Sciences, 1971.

[2] CSTR/SSB Study on the IMS, National Academy of Sciences, 1973.

[3] The Outer Solar System, A Program for Exploration, National Academy of Sciences, 1969 (see recommendations 5 and 6 on p. 9).

[4] Outer Planets Exploration, 1972-1985, National Academy of Sciences, 1971 (see recommendation j on p. 23).

[5] Proceedings of the Jupiter Radiation Belt Workshop, edited by A. J. Beck; also JPL Tech. Memo. 33-543, July 1972.

[6] Scarf, F. L., "Characteristics of the Solar Wind near the Orbit of Jupiter," Planetary and Space Sciences, Vol. 17, 1969, p. 595.

[7] Scarf, F. L., "Some Comments on the Magnetosphere and Plasma Environment of Saturn," Cosmic Electrodynamics, Vol. 3, 1973, p. 437.

[8] Coroniti, F. V., Kennel, C. F., and Thorne, R. M., "A Model for Jovian Electron and Proton Fluxes," EOS, Vol. 54, 1973, p. 446.

[9] Brice, N. M., and McDonough, T. R., "Jupiter's Radiation Belts," Icarus, Vol. 18, 1973, p. 206.

[10] Michel, F. C., and Sturrock, P. A., "Centrifugal Instability of the Jovian Magnetosphere and its Interaction with the Solar Wind," EOS, Vol. 54, 1973, p. 445.

[11] Birmingham, T. J., Hess, W. N., and Northrop, T. G., "The Electron Radial Diffusion Coefficient in Jupiter's Magnetosphere," EOS, Vol. 54, 1973, p. 446.

[12]Piddington, J. H. and Drake, J. F., "Electrodynamic Effect of Jupiter's Satellite, Io," Nature, Vol. 217, 1968, p. 938.

[13]Goldreich, P. and Lynden-Bell, D., "Io, a Jovian Unipolar Inductor," Astrophysical Journal, Vol. 156, No. 1, Part 1, April 1969, pp. 59-78.

[14]Spitzer, L., Jr. and Harm, R., "Transport Phenomena in a Completely Ionized Gas," The Physical Review, Vol. 89, 1953, p. 997.

[15]Gurnett, D. A., "VLF Hiss and Related Plasma Observations in the Polar Ionosphere," Astrophysical Journal, Vol. 175, No. 1, Part 1, 1972, pp. 525-533.

[16]Fredricks, R. W., Scarf, F. L., and Russell, C. T., "Field-Aligned Currents, Plasma Waves, and Anomalous Resistivity in the Disturbed Polar Cusp," Journal of Geophysical Research, Vol. 78, No. 13, 1973, pp. 2133-2141.

[17]Gurnett, D. A. and Frank, L. A., "VLF Hiss and Related Plasma Observations in the Polar Ionosphere," Journal of Geophysical Research, Vol. 77, No. 1, 1972, pp. 172-190.

[18]Scarf, F. L., Fredricks, R. W., Green, I. M., and Russell, C. T., "Plasma Waves in the Dayside Polar Cusp, Part 1: Magnetospheric Observations," Journal of Geophysical Research, Vol. 77, No. 13, 1972, pp. 2274-2293.

[19]Wu, C. S., "Triggering Mechanism for Io-Modulated Decametric Radio Emissions from Jupiter," EOS, Vol. 54, 1973, p. 446.

[20]DeForest, S. E., "Spacecraft Charging at Synchronous Orbit," Journal of Geophysical Research, Vol. 77, No. 4, 1972, pp. 651-659.

[21]Fredricks, R. W. and Scarf, F. L., "Observations of Spacecraft Charging Effects in Energetic Plasma Regions," Photon and Particle Interactions with Surfaces in Space, edited by R. J. L. Grard, D. Reidel Publishing Company, Dordrecht-Holland, 1973, pp. 277-308.

[22]Warwick, J. W., "Decay of Spin in Sputnick 1," Planetary and Space Sciences, Vol. 1, No. 1., 1959, pp. 43-49.

For comments on the Pioneer 10 encounter results see next page

Comments on the Pioneer 10 Encounter Results
(Note added in February 1974)

The successful Pioneer 10 flyby of the Jupiter magnetosphere in December 1973 provided direct first order information
on the planetary magnetic dipole moment, the apparent trapped
radiation population within 20 R_J, and the overall configuration of the magnetosphere out to about 100 R_J. In addition,
although Pioneer carried no plasma physics instrumentation,
the observations from the other instruments show conclusively
that the unmeasured plasma physics phenomena actually control
the entire magnetosphere. Finally, some Pioneer experimenters
pointed out that the absence of plasma diagnostics leads to an
enormous uncertainty in analysis of the trapped radiation measurements, and a corresponding ambiguity can arise in interpreting the ionospheric profiles.

The basic phenomena revealed by the Pioneer 10 observations can be summarized as follows (see Science, 183, 301-325,
1974):

    a. The magnetic dipole moment is only 4 gauss-R_J^3, rather
than the (10-12) gauss-R_J^3 value previously estimated, but the
orientation and offset are similar to the values deduced by
radio astronomers.

    b. Despite the small value of the surface field, the
Jupiter magnetosphere is much larger than anticipated, because
an unmeasured "thermal" plasma drags the field outward and
causes it to be distorted into a sun-like radial spiral. The
$\beta \simeq$ (1-4) plasma may involve energized photo-electrons or
secondaries from the Jovian atmosphere, and the variable interaction with the solar wind (over the range 50 to 100 R_J) may
involve two-stream instabilities, as conjectured by Michel and
Sturrock.[10] The planetary field lines can merge with the
interplanetary field over the entire outer region, with
current-driven plasma instabilities providing the dissipation
mechanism.

    c. Within about 20 R_J, the trapped energetic electron
levels are up to a million times greater than those found on
Earth, and up to a hundred times greater than predicted by the
Radiation Belt Workshop upper limit model. In fact, the peak
fluxes at several Mev are similar to those shown in the center
panel of Fig. 3.

d. The trapped proton fluxes fall off strongly within
$L \simeq 3.6$, suggesting that an electrostatic or electromagnetic
ion cyclotron instability is very effective at low L-values.

e. There is evidence for collisionless local acceleration (to Mev energies) of electrons and protons throughout the
magnetosphere out to the bow shock, and intense fluxes of
usually energetic upstream particles were observed over vast
distances.

f. The Pioneer 10 spacecraft could have charged to Mega-
volts within 10 R_J; the measured fluxes in this region were
comparable to expected photoemission fluxes, and the apparent
flux of lower energy electrons (0.1 to 2 Mev) dropped off
within about 10 R_J. In fact, some spacecraft anomalies and
false commands were detected in this region, and these could
be attributed to impulses associated with sheath fluctuations.
However, without an unambiguous sheath-independent measurement
of N(thermal) (such as that provided by detection of the
plasma frequency wave cutoff or lower hybrid resonance wave
emission) one cannot determine the sheath conditions or the
magnitude of the sheath correction at Mev energies.

g. The origin of the Jupiter ionosphere will also be in
doubt because the observed trapped particle fluxes are so
close to the expected stable trapping limits [for nominal
N(thermal)-values] that the precipitating particle flux may
well control the ionization, as it does in the Earth's polar
ionosphere. It is certainly true that the trapped radiation
level at Jupiter is a million times more intense than at Earth
while the UV flux at Jupiter is 1/27 that of Earth, so that
this auroral or polar ionosphere analog is actually a very
likely one. In fact, since the peak ionospheric electron density at Io is within a factor of two of the Venus value, while
the UV flux is down by a factor of 50, it seems that UV cannot
be the dominant ionization source near Jupiter.

h. Since the observed dipole field strength is so low,
it is difficult (if not impossible) to explain the decametric
radiation in terms of ionospheric gyrofrequency radiation.
However, if magnetospheric wave-particle interactions provide
enough precipitation to enhance the ionospheric density above
about $5 \times 10^6 \mathrm{cm}^{-3}$, electrostatic emissions at $(n + 1/2)\ f_c^e$
(see Fig. 1) can couple strongly to the radiation field and
account for the observed decametric spectrum.

For "Note Added in Proof" see following page

Note Added in Proof

The section containing additional comments on the Pioneer 10 encounter observations was prepared early in 1974, and later in the same year, Pioneer 11 successfully traversed the Jupiter magnetosphere. The initial Pioneer 10 post-encounter comments still give a reasonable summary description of the basic new phenomena revealed by all the in-situ observations, but, of course, in many areas more comprehensive analyses and interpretations have already appeared in journal articles and technical books. In addition, for the problem area involving charging of the spacecraft (point f, above), some significant information comes from combining the Pioneer 10 and 11 observations. Since the details have not yet been discussed in print, the relevant encounter data and speculations are briefly summarized here.

The phenomenon of interest is related to the detection of spacecraft anomalies on Pioneer 10/11. Similar anomalies are detected in synchronous orbit at Earth when substorm plasma injections lead to rapid variations in spacecraft potential. In order to determine if the Pioneer 10/11 anomalies occurred when the spacecraft traversed surfaces associated with uniform plasma conditions, it is first necessary to plot the encounter trajectories in magnetic coordinates, and such a plot is shown in Fig. 9, where the best fit (D_2) magnetic field model (based on Pioneer 10 data analysis) is used. In these magnetic coordinates, the various L-shell contours are associated with distinct flux values for the trapped energetic particles, and the contours illustrated in Fig. 9 are labeled in terms of the measured omni-directional fluxes for E > 35 Mev electrons [determined by the Pioneer 10 Trapped Radiation Detector (TRD) of Fillius and McIlwain].

Fig. 9 shows that on Pioneer 10 and 11 a substantial number of anomalies were detected near L ≃ 12-13, or just in the region where current-balance considerations would suggest a sheath reversal. On Pioneer 10 these anomalies included spurious commands for the photopolarimeter (IPP) changes in the level of the spacecraft receiver (AGC) and commutator anomalies for the Trapped Radiation Detector (TRD). On Pioneer 11, spacecraft heaters and the conscan mode were spontaneously turned on at the same magnetic L-shells.

Although there is no direct proof at all that fluctuations in spacecraft potential induced these anomalies, it is noteworthy that the L ≃ 12-13 shell is just where the measured energetic electron current density (E > 160 keV) would be

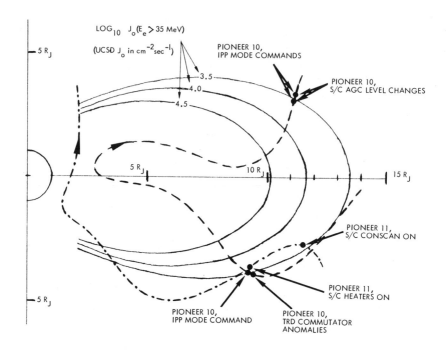

Fig. 9   Locations of anomalies along Pioneer 10,11
          trajectories. D_2-model magnetic coordinates.

approximately equal to the current per unit area associated
with a Brice-type cold plasma distribution. Thus, it is
plausible that there were large fluctuations in local spacecraft potential at the times of these spacecraft anomalies,
and such fluctuating signals could have been picked up on
the pins of the unshielded spacecraft test connector. Thus,
in principle, there is a simple way in which external sheath
fluctuation effects could have induced spacecraft anomalies
of the type shown.

SELECTION OF PIONEER 11 TARGET POINT

Fred D. Kochendorfer*

NASA Headquarters, Washington, D. C.

Abstract

One of the primary objectives of the Pioneer 10 and Pioneer 11 missions was to explore the Jovian environment. The initial targeting point of Pioneer 11 had been selected to provide the greatest number of targeting options so that the specific selection of the Pioneer 11 flyby trajectory could await the results from the Pioneer 10 flyby. Following the extremely successful Pioneer 10 flyby of Jupiter, a final decision had to be made on the flyby trajectory for Pioneer 11, since some of the targeting options required a midcourse maneuver of the Pioneer 11 spacecraft only a few weeks after the Pioneer 10 flyby. Based on the quick-look analyses of the Pioneer 10 data, it was decided that the data on the Jovian environment which would best complement and extend the excellent results from Pioneer 10 would be acquired from a target point at Jupiter that would provide a close, high-latitude (45°), left-side passage of the planet. One such targeting option would take the spacecraft on to Saturn. It was recommended that Pioneer 11 be targeted at this "Saturn point." Some of the logic that was used in arriving at this decision is summarized below.

Magnetic Fields

The properties of the magnetic field, the tilt and offset of the dipole and the nature of higher-order poles, are important not only in their own right but also are vital to our understanding of the radiation-belt data. For example, higher-order poles, if they exist, could result in lower fluxes of high-energy particles close to the planet.

---

* Pioneer Program Manager.

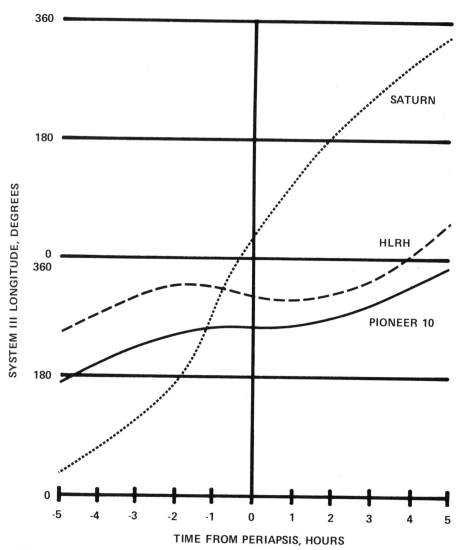

Fig. 1   Comparative Ranges of Longitudes within 5 hr
          of Periapsis for Pioneer 11 Options and Pioneer 10.

To improve upon the model from the Pioneer 10 data, it is
necessary 1) to cover a wider range of longitude while in
close to the planet in order to reduce the uncertainty in
dipole offset and tilt; and 2) to go closer to evaluate the
higher-order poles. It can be seen from Fig. 1 that, during
the period about 5 hr from periapsis ($R < 7$ R_J): Pioneer 10
covered a longitude range of approximately 160°; a high-

latitude right-hand (HLRH) passage produces about the same range; but a left-side passage like the "Saturn point" produces an excellent range of 650°. For measurements of fields, then, a closer left-hand passage is desired.

Radiation Belt

For radiation-belt measurements, possibilities for extending the Pioneer 10 results are illustrated in Fig. 2, which shows several trajectories in coordinates relative to the magnetic field model. Pioneer 10 traversed a region between ± 20° lat up to its closest-approach distance of 2.8 R_J. Two features of the Pioneer 10 data are of special interest and will be given further elaboration. The data show a strong latitude effect, so that an extension to a wider range of latitudes is desired. In addition, the 30-Mev proton data show a tantalizing drop in particle count starting 1 1/2 hr before periapsis at about 3.4 R_J; the counts drop by an order of magnitude into periapsis (2.8 R_J) and then rise

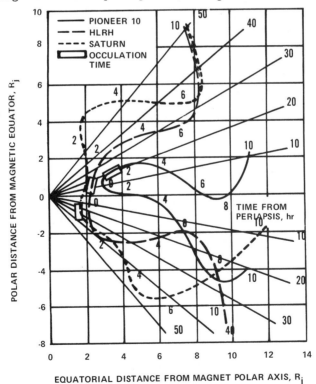

Fig. 2   Pioneer trajectories in Jovian magnetic coordinate system.

almost to the inbound peak about 1 hr later, again at about
3.4 R_J. It is important, both for an understanding of the
inner belt and for an evaluation of the potential of future
orbiter missions, to determine whether Pioneer 10 measured a
local disturbance between 3.4 and 2.8 R_J or whether the trend
continues inward. It is possible to cover a larger latitude
range by selecting a high-latitude aim point on either side.
The left side, however, is somewhat better. As shown in
Fig. 2, latitudes from -40° to +65° are traversed at dipole
axis distances less than 2 R_J. The high-latitude aim point
has another important feature: the spacecraft is in the high-
radiation region near the equator for relatively short periods
of time, and it will be possible to target closer without exceeding the total fluence absorbed by Pioneer 10. Thus for
extension of the Pioneer 10 radiation-belt data, a closer,
high-latitude target point is desired with a slight preference
for the left side.

Imaging

The Principal Investigator for the Imaging Photopolarimeter (IPP) has stated that, to best complement the Pioneer
10 data, a "polar" target point should be selected, where
"polar" means any point having a latitude greater than 45°.
Picture quality, in terms of overlap or underlap between
successive scan areas, will be about the same for any approach
trajectory up to a distance of about 5 R_J. In closer, the
underlap is greater for a left-side trajectory as illustrated
by a comparison of the curve for the Saturn point with that
for HLRH in Fig. 3. If the image is centered at the polar
region, instead of the subspacecraft point, however, the
underlap is independent of spacecraft direction around the
planet. The IPP thus desires high latitude with a slight
preference for the right side.

Infrared Radiometer

The argument is the same as for imaging except that
underlap is not a disadvantage because it is offset by a
larger area of coverage. For the Saturn point two view periods
occur, and for HLRH only one occurs. High latitude is desired
with slight preference for the left side.

Target Point Summary for Best New Data on Jupiter

The experiments not covered previously have no preference
for the right or left side. The uv does not see Jupiter for

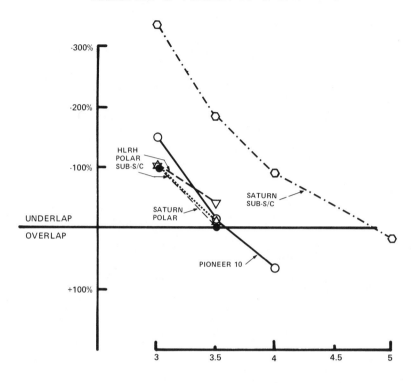

Fig. 3   The amount of underlap of imaging picture elements
from the IPP on Pioneer 11 for Pioneer 10 and Saturn
and HLRH options.

any high-latitude pass, but the experimenter sees no reason
to repeat the Pioneer 10 measurement on the planet, and a
repeat of the neutral hydrogen ring and satellite measurements
can be made from either side. As a result, the target region
that best complements and extends the overall scientific
results from Pioneer 10 is closer, high-latitude, and left-
hand. In a meeting on December 13, 1973, the consensus of
the experimenters was strongly in favor of the Saturn point.

Occultation

In Fig. 2, both high-latitude trajectories provide an
occultation of Jupiter. For the HLRH option, occultation
starts 8 min after crossing the magnetic equator and continues
for about 30 min. For the Saturn point, occultation lasts
44 min and ends 8 min before crossing the magnetic equator.
During occultation, the spacecraft covers latitude range from
about −40° to −10°, and since this represents a period of

importance it will be necessary to insure that Pioneer 11
survives occultation. This will be discussed further.

Effects of Jovian Radiation on Pioneer 10

During passage through the intense radiation environment,
three types of effects were observed, as described in the
following paragraphs.

Permanent Damage

The optics of both the stellar reference assembly and
the asteroid-meteoroid detector were darkened by at least 10%.
Both, however, are still working. A few components (probably
transistors) have failed in the data analysis circuits of the
cosmic ray telescope (CRT), resulting in a slight degradation
in the ability to readout data. Since the CRT electronics
are outside of the equipment compartment, they can "see"
electrons having energies in excess of 1 Mev. Although similar devices have survived ground testing at fluences 10 times
that seen by Pioneer 10, failure of a few "mavericks" out of
the total of over 50,000 in the box probably is not surprising.

Temporary Damage

Slight changes were observed in the telemetered values
for power system current and voltage, transmitter power ampli-
fier current, and oscillator frequency. The effects were
observed just before periapsis and persisted for several days.
They caused no problems, and all of them have returned to
nearly their values before encounter. A number of the measurement are made by zener diodes, and it is probable that the
diodes changed rather than the actual currents.

Temporary Anomalies

"Uncommanded" changes in operating mode of the IPP occurred during a period starting just before periapsis and
continuing for about 40 hr. It has not been determined
whether these were due to radiation-induced noise or to some
combination of noise and a very heavy command activity. In
general, the Pioneer electronics were designed for immunity to
noise from electron fluxes of at least two orders of magnitude
above those observed. No uncommanded changes were observed
in other equipment. For Pioneer 10, the loss of data due to
uncommanded changes was minimized by sending frequent commands
that insured proper mode for the instrument (and the space-

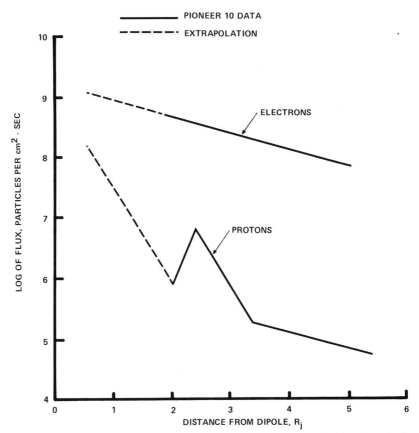

Fig. 4   Radiation Workshop model of Jovian radiation belt.
          Flux along magnetic equator from Pioneer 10 data.

craft). A similar procedure will be used for Pioneer 11.
Despite the fact that Pioneer 10 survived in relatively good
condition, it is probably unreasonable to select for Pioneer
11 any important objectives that require survival of a total
fluence much in excess of that seen by Pioneer 10.

Pioneer 10 Radiation Model

The trajectories of interest for Pioneer 11 cover ranges
of latitude and radius from Jupiter which require an extrapolation of the region covered by Pioneer 10. However, the
extrapolation requires a model that represents Pioneer 10
data. In mid February, to support studies of the best target
point for Pioneer 11, representatives of all field and particles experiments on Pioneer 10 produced a workshop model that
is in good agreement with all data. This model and a conservative extrapolation are shown in Fig. 4.

Table 1 Calculated fluxes and fluences

Radiation model

	Total fluence[a]		Fluence[a] at exit from occultation		Peak flux[b]	
	E	P	E	P	E	P
Pioneer 10	660	3.2	...	...	5×10^8	7×10^6
Saturn Point	150	7.9	79	4.5	1×10^9	1.2×10^8

[a]Fluences in 10^{10} particles/cm^2; E ~ electrons > 3 Mev; P ~ protons > 30 Mev

[b]Fluxes in particles/cm^2/sec; E ~ electrons 73 Mev; P ~ protons > 30 Mev.

Calculated Radiation Fluences

Electron and proton fluences calculated for the workshop model are shown in Table 1. For the Saturn point, the total fluence, as well as the fluence up to exit from occultation, is shown. A comparison with Pioneer 10 values shows that electrons are no problem. For protons, the very conservative extrapolation shown in Fig. 4 produces a total fluence of about twice that for Pioneer 10. However, since the fluence at occultation exit is about the same as for Pioneer 10, there is high confidence in receiving the important data stored during occultation. For fluxes, peak values for the Saturn point are three to ten times those seen by Pioneer 10 but are below those that should cause noise problems in the circuits.

Recommended Target Point

The Pioneer Project Office and the Program Office recommended a Pioneer 11 target at the Saturn point because the resulting data at Jupiter should best complement and extend the excellent results from Pioneer 10. Based on an extrapolation of the Pioneer 10 data which is believed to be

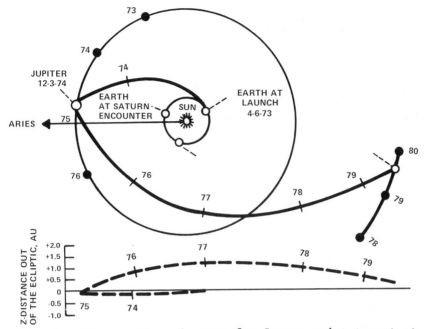

Fig. 5   Pioneer 11 trajectory for Saturn point target at
         Jupiter.

conservative, Pioneer 11 should survive the encounter on this
trajectory.

Post-Jupiter Trajectory

One of the bonus features of the Saturn point is that
Pioneer 11 will leave Jupiter in a direction toward the sun
and reach a perihelion, just inside the asteroid belt, at
3.5 a.u. early in 1976 (Fig. 5). The orbit plane will be
inclined 16.5° to the ecliptic, and, although not an "out-of-
ecliptic" mission, it will be the highest inclination of any
spacecraft up to that time.

Pioneer 11 Capabilities at Saturn

In September 1979, when Pioneer 11 arrives at Saturn,
there will be a power deficiency of about 8 w, if the degradation rate of the radioisotope thermoelectric generators
(RTG's) does not increase. Operation of all experiments
requires 33 w. However, since the asteroid-meteoroid experiment will probably be off and the infrared radiometer can be

off except for the few hours of its view period, 6 w will be
saved, and only one additional instrument at any one time will
need cycling.

With a 64-m Deep Space Net (DSN) station, the communi-
cations systems will have an estimated -0.4-dB margin at 512
bits/sec (which will result in a small frame deletion rate)
or a +2.6 dB margin at 256 bits/sec. As a result, a large
quantity of data can be returned.

The scientific results on Saturn will depend greatly on
the target point, and careful consideration of factors such
as the best point to assist Mariner Jupiter/Saturn (MJS) and
the best point to extend MJS data, etc., must occur before
the target can be selected. Therefore, only a general capability for science at Saturn is discussed below.

Little is known about the particle and field environment
near Saturn from Earth-based measurements, since, unlike
Jupiter, no rf noise from Saturn has been observed. The
possibility of radiation belts at Saturn cannot be discounted,
however, since the greater distance of Saturn from Earth
would result in the strength of noise signals from Saturn
being only about one-quarter of those from Jupiter, even if
the Saturn source strength were the same as that of Jupiter.
Thus, Pioneer 11 will be capable of mapping exploratory magnetic fields and particle fluxes at Saturn in the same manner
as at Jupiter. The presence or absence of a magnetic field
and a radiation belt would be a new discovery and probably
would have a greater impact on scientific ideas about the
outer planets than would the similar measurements at Jupiter
which were more of a refinement of Earth-based measurements.

Ultraviolet measurements at Saturn are possible in
Pioneer 11. Although the intensity of the reflected light is
smaller at Saturn than at Jupiter, inquiries of the principal
investigator showed that the sensitivity of the instrument is
adequate to make the measurements. The presence or absence
of helium and perhaps the determination of the hydrogen/helium
ratio for Saturn would be new discoveries.

Infrared measurements at Saturn are possible on Pioneer
11; the infrared radiometer (IRR) instrument is sufficiently
sensitive to perform the measurements. Establishment of the
thermal balance of the planet would be the result of measurements similar to those performed at Jupiter.

Imaging at Saturn should be even more spectacular than
at Jupiter, primarily because the resolution of Earth-based

pictures of Saturn is only about one-quarter of those of
Jupiter, whereas the resolution of the IPP instrument is the
same in both cases. Although the light intensity at Saturn
is only about one-quarter of that at Jupiter, the sensitivity
of the instrument is more than adequate to make the measurement. (The gain of the instrument can be increased in nine
steps by ground command to about 80 times that used at Jupiter
by the Pioneer 10 instrument.)

Photometry and polarization measurements of the reflected
light from Saturn at phase angles impossible from Earth-based
measurements are possible also on Pioneer 11 using the IPP
instrument. Such measurements should provide new information
about the characteristics of the Saturn rings and composition
of any Saturn atmosphere.

Measurement of dust clouds in the vicinity of Saturn by
the meteoroid detector is possible also on Pioneer 11. This
capability on Pioneer 11 at Saturn will be poorer than on
Pioneer 10 at Jupiter, however, because the sensors on Pioneer
11 are twice as thick as those on Pioneer 10 and because the
remaining useable penetration area on Pioneer 11 at Saturn

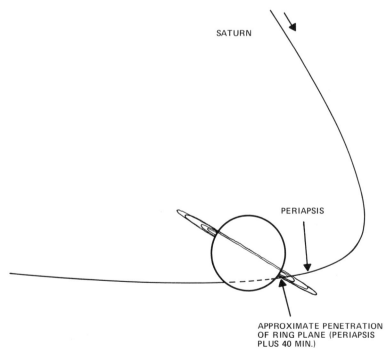

Fig. 6   Typical trajectory as seen from Earth.
Outside rings.   RCA ≈ 2.25 R$_S$; θ = 0°

SATURN

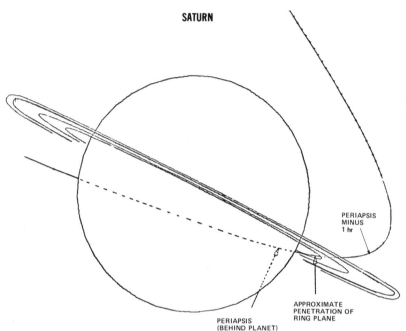

PERIAPSIS
MINUS
1 hr

APPROXIMATE
PENETRATION OF
RING PLANE

PERIAPSIS
(BEHIND PLANET)

Fig. 7   Typical trajectory as seen from Earth.   Inside inner
ring.   RCA=1.15; R_S; θ = 22°

may be less than that on Pioneer 10 at Jupiter, since the
flight time to Saturn is almost 4 times that to Jupiter.   The
mass of Saturn also can be determined more precisely than from
Earth measurements by tracking the spacecraft as it goes past
Saturn.   Determination of the structure and density of the
Saturn atmosphere and perhaps that of the rings also is
possible if the spacecraft is occulted by Saturn and its rings.

    Examples of the two types of Saturn flyby trajectories as
seen from Earth are shown in Figs. 6 and 7.   In Fig. 6, the
radius of closest approach (RCA) is 2.25 R_S, just outside of
the outer ring.   The dots on the figure are 15 min   apart.
The spacecraft is occulted by all of the rings, then passes
into the clear for over 1/2 hr, and then is occulted by Saturn.
Good view angles of Saturn and the rings should be available
for all instruments.

    In Fig. 7, the RCA is 1.15 R_S, and the spacecraft passes
between the inner ring and the planet, a point that MJS will
not cover but of potential interest for an orbiter.   Ring
occultation occurs, followed by a 1/2-hr clear period as the
spacecraft flies through the "gap," and then Saturn occultation.   Again, good views should be available.

CHAPTER III—COMETS

SOME RECENT DEVELOPMENTS
IN COMETARY PHYSICS

D. A. Mendis [*]

Department of Applied Physics and Information Science,
University of California, San Diego; La Jolla, Calif.

Abstract

Recent observations and associated theoretical developments bearing on the composition and structure of comets are briefly reviewed. The physical and dynamical processes in cometary atmospheres and ionospheres are discussed, and detailed hydrodynamic models of expanding multiconstituent cometary atmospheres corresponding both to a central nucleus as well as a central nucleus surrounded by a distributed source are presented for comparison with observation. It is argued that observed "slow" speed for the neutral hydrogen is not incompatible with a purely H_2O "parent" source for that species.

Introduction

Although comets, for the most part, still remain rather mysterious members of the solar system, several devel-

Presented as Paper 73-549 at the AIAA/AGU Space Science Conference: Exploration of the Outer Solar System, Denver, Colorado, July 10-12, 1973. Performed under contract NGR-05-009-110 issued by the Planetology Program Office, Office of Space Sciences, National Aeronautics and Space Administration.
[*]Associate Research Physicist and Lecturer.

opments within the last three or four years have contributed
very strongly to our present understanding of them. In this
paper an attempt is made to assess a few of these developments, although the limits set by the author's own range of
interests will no doubt tend to make the total picture somewhat lacking in proper perspective.

The Observational Features of a Comet

A typical comet when sufficiently close to the sun exhibits three essential features: a coma, a nucleus, and a tail
(type I or type II or both).

The coma is a diffuse luminous region approximately
spherical in shape whose visible boundary merges with the
sky-background. In the optical region, in which it was exclusively observed until recently, it is seen by the emission
bands of various radicals, a few atomic lines including the
forbidden (red) lines of neutral oxygen and also emission
bands of ions occurring in the tails. It further shows the reflected Fraunhofer spectrum of the sun indicating the presence of solid bodies in the form of dust or larger chunks.

More recently two long period comets Tago-Sata-
Kosaka [1969g] and Bennett [1969i] as well as the short period
comet Encke has been observed by the u-v detectors on the
orbitors OAO-2 and OGO-5. They have all shown extensive
envelopes of strong Ly-α emission. [1] Several comets have
also been observed recently in the infra-red. They show a
strong thermal component. [2] The spectral identifications in
comets are shown in Fig. 1. The existence of these metallic
species as well as scandium in comets is supported by evidence on meteor streams. [3] When certain meteor streams,
notably the β-Taurids (which is supposedly associated with
comet P/Encke and the Leonids (which is almost certainly
associated with P/Swift-Tuttle), intersect the earth's upper
atmosphere there are significant enhancements of the ions of
all these metals.

The size of the coma of course depends firstly on the distance from the sun and secondly on the particular emission used. In the optical region the (o-o) rotation-vibration band of CN in the blue ($\lambda \approx 4000$Å) is the strongest spectral feature. It is also the first to appear ($r \approx 3$ A.U.) and defines the greatest extension of the head. The [01] lines appear when $r \lesssim 1$ A.U.

HEAD:	CN, C_2, C_3, CH, $C^{12}C^{13}$, NH, NH_2,
	[01], OH, Na, Si, Ca, Cr, Mn, Fe,
	Ni, Cu, K, Co, Y.
	H (ULTRAVIOLET)
	CO^+, CH^+, CO_2^+, N_2^+, OH^+, Ca^+
	REFLECTED SUNLIGHT
	THERMAL EMISSION (INFRARED)
TAIL (TYPE 1):	CO^+, CH^+, CO_2^+, N_2^+, OH^+.
TAIL (TYPE II):	REFLECTED SUNLIGHT
	THERMAL EMISSION (INFRARED).

Fig. 1 Spectral identifications in comets (see addendum)

typically, whereas the metallic lines appear only when $r \lesssim$ 0.1 A.U. and have so far been identified only in a handful of cases.

The recent U-V observations have shown that the greatest extension of the coma is in the Ly-α emission of neutral hydrogen. Comet Tago-Sato-Kosaka [1969g] when at a heliocentric distance of 1 A.U. showed a Ly-α coma of about 10^6 km and comet Bennett [1969i] at the same distance showed one ten times larger.

The most spectacular feature associated with a comet is its plasma (type I) tail, which when fully developed extends 20-30 million km. Normally the tail begins to develop when $r < 1.5$ A.U. although cases are known when it appeared much earlier. The best known example is comet Humason which appeared to develop a type I tail when it was still about 5 A.U. from the sun.[4] The strongest emission is from CO^+ (the other emissions being shown in Fig. 1). The dust (type II) tail often separates out of the gas tail and lags behind the latter which points almost radially away from the sun. The plasma tails show considerable structure, e.g. rays, knots, helical features, sheets, etc., which seem to indicate the presence of magnetic fields. The dust tails, in contrast, show practially no structure.

Velocities and accelerations of the cloud like condensations (knots) have been calculated: the velocities range from 10 to 300 km sec^{-1}, while the accelerations range from 100-1000 cm sec^{-2}. The acceleration, which in units of solar gravity at the point assumes values typically around 100 cannot be explained in terms of radiation pressure, which is due to the resonant scattering of solar radiation on various lines. In fact the radiation force is typically less than 1/10 of the gravitational force. One has to look for another mechanism.

It is clear that the solar wind with its frozen-in magnetic field must play a dominant role in sweeping the ionized components of the coma into the tail and also in shaping and maintaining the tail as it streams away in the anti-solar direction.[5] There is sufficient momentum in the solar wind and adequate coupling between it and the cometary ionosphere via the embedded magnetic field to be ultimately responsible for the acceleration observed in the tail. What is less clear is the manner in which the interplanetary magnetic field is mixed with the coma plasma in such a way as to produce the observed fine structure in the tail.

The solar wind-comet interaction is shown schematically in Fig. 2. The interplanetary magnetic field convected

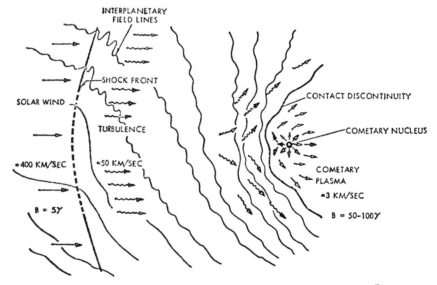

Fig. 2   The comet-solar wind interaction.[5]

by the solar-wind cannot diffuse through the cometary ionosphere in a time comparable with the time it takes to flow
past. Consequently it piles up against the cometary ionosphere being separated from it by a contact discontinuity
(magnetosheath). The contact surface would typically be at a
distance of 10^5 km from the center and the enhanced magnetic field at the stagnation point around 50-100 γ. The solar-
wind being super-magnetosonic and super-Alfvénic must pre-
pare well upstream for the encounter with the ionized coma
by decelerating via a collisionless shock (like the earth's
bow shock) or via a transonic ion exchange sheet. Biermann
et al[6] believe in a shock typically 5×10^6 km upstream from
the nucleus, whereas Wallis[7] proposes the transonic process
with no shock. The basis of the transonic process is that the
incoming solar plasma loses momentum as it gradually
picks up heavy or slow moving cometary ions ahead of the
contact surface, consequently it might go smoothly from
supersonic to subsonic flow. It is difficult to choose between
the two models at this stage because Wallis' model is one-
dimensional (it neglects flow divergence altogether) while
Biermann's model is quasi-one dimensional, allowing for the
flow divergence only in an ad hoc manner. However, should
a shock exist it would be considerably weakened. In fact, the
most recent calculations of Wallis[8], with regard to comets
Bennett and Tago-Sato-Kosaka does suggest a weak shock at
a distance of $2-3 \times 10^5$ km from the nucleus. Both the build
up of ions inside the contact surface caused by their being
pushed against it by the outflowing neutrals, and the pressure
balance across this surface needs further investigation. It
seems possible that the contact surface is more like 10^4 km
from the nucleus since plasma number densities of the order
of 5×10^4 cm^{-3} would be required to produce the required
pressure. At such densities dissociative recombinations of
molecular ions seem likely (e. g., CO^+ has a lifetime against
dissociative recombination of about 10^2 - 10^3 sec at such
densities [9], and these times are less than the flow times
across the coma with velocities around 1 km sec^{-1}).

It has been pointed out that the contact surface is liable
to flute instabilities because the magnetic field is curved in
such a way that it is likely to enter the coma plasma on con-
tracting, and that this may be the way in which the inter-

planetary field mixes with the plasma in the tail[10] but the
process needs to be investigated in detail.

As regards the type II dust tails, they appear to be
composed of particles around micron size. (The reflected
solar spectrum being slightly reddened.) The forms of these
tails may be explained by the Bessel-Bredichin theory where-
in the dust particles move freely in the combined gravitational and radiative fields of the sun. Although this model
seems to be adequate in explaining pure type II tails, when
type I and II tails are present together there is obviously a
strong gas-dust interaction near the head as is indicated by
the near radial orientations of the dust tails in this region.
Finson and Probstein[11] have modified the Bessel-Bredichin
theory by taking into account the drag exerted by the gas on
the dust in the inner region. The dust appears to reach terminal velocities of about 0. 3 km sec^{-1} at a distance of about
100 km from the nucleus before it is decoupled from the outflowing gas.

The most controversial component with regard to its
nature is the nucleus. It is never seen with the naked eye.
With large telescopes it has an almost starlike appearance at
the center of the coma. In some comets nuclei cannot be observed whereas in others multiple nuclei are observed. Also
occasional splitting of nuclei, as in the well known case of
comet Beila, have been observed. Even when no nucleus is
observed one cannot reach an unambiguous conclusion about
its existence or non-existence. The fractional contribution
of the nucleus to the integrated brightness of the coma is
typically less than one percent. So only big telescopes with
large magnification succeed in separating the starlike nucleus
from the coma. From the lack of resolution of the nucleus
coupled with the maximum resolution of the telescopes one
obtains upper limits for the radius of about 100 km. One
can, however, proceed to estimate the size from the observed brightness coupled with some assumptions about the
albedo and phase correction. Assuming the smallest albedo
in the solar system (≈ 0.02) one obtains, according to
Roemer[12] the following values:

Short period comets: 0. 8 km < r < 38 km

Long period comets: 2 km < r < 65 km.

If an albedo of 0. 7 (corresponding to the largest in the
solar system) is assumed all the above values are decreased
by a factor of about 6. The difference in the size of long and
short period comets is clearly significant but it is not entirely clear whether this is intrinsic or is merely a selection
effect.

If the assumption is made that the nucleus is a single
monolith having meteoritic bulk densities one comes out with
the masses in the range 10^{16} - 10^{23} grams. Lack of observational gravitational effects (e. g., comet P/Brooks 2 passed
through the satellite system of Jupiter in 1886 without causing
any noticeable effect) indicate upper limits of about 10^{20} gms.
Lower limits may be derived from the observed rate of loss
of gas and dust which is typically about 10^{13} - 10^{14} gms per
revolution (e. g., Arend-Roland, Mrkos, etc.).

As regards the nature of the nucleus, although a minor-
ity view is that the nuclear region is a "flying gravel bank"
having no physical or gravitational coherence[13], the majority
opinion holds that it is a monolith. The generally accepted
model is some variant of Whipple's "icy conglomorate"
model[14] which asserts that the nucleus consists of a matrix
of frozen ices and meteoric dust. This model has been successful in explaining, both qualitatively and quantitatively a
variety of cometary phenomena such as the nongravitational
effects, sudden breakups, flares and also the general
features of the expanding coma. Spiral shaped jets which
have been occasionally observed visually in the inner coma
and which have been photographed in the case of comet
Bennett[15] also seem to support the existence of a central
nucleus which is apparently rotating with a period of the
order of a day.

The Composition of the Nucleus

While the "icy conglomerate" model of the cometary
nucleus has been successful in explaining a variety of cometary phenomena as described earlier, one needs to know
its chemical composition (especially that of its volatile component) in order to interpret the activity of the coma.

The chemical instability of the radicals observed in the coma suggested that these could not be stored in the nucleus for sufficiently long times and were likely to be the photo-dissociation products of more chemically stable "parent molecules" such as H_2O, NH_3, CH_4, etc. [16] Delsemme and Swings [16] had also suggested over twenty years ago that these parent molecules may be present in the nucleus as clathrate hydrates. Clathrate hydrates of gases (loosely called gas hydrates) are formed by a peculiar lattice of H_2O ice containing cavities where many types of gas molecules may be encaged by van der Waal's forces. Since the potential wells in which these "guest molecules" are trapped are very deep they can be released only by the destruction of the "host" H_2O lattice and consequently their vaporization is controlled by the latent heat of vaporization of H_2O. This beautifully explains the almost simultaneous appearance of all the major cometary emission bands (typically around 3 A. U. for most comets) although the volatilities of the assumed parent molecules differ by over ten orders of magnitude (see Fig. 3). Miller [17] as well as Delsemme and Wenger [18] have also pointed out that under typical cometary conditions the clathrate is thermodynamically more stable than its constituents. It is worth noticing that the radicals themselves (rather than their "parent molecules") may be stored in the nucleus in this fasion, because they will be held in "deep-freeze" in the deep potential wells away from their neighbors. About 17% (by number)

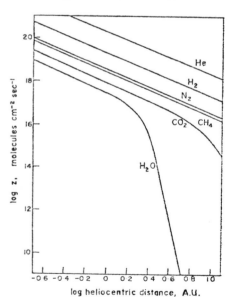

Fig. 3 Vaporization rate Z, mol cm^{-2} sec^{-1}, for various snows as a function of heliocentric distance, in A. U., computed for the steady state temperature of a rotating cometary nucleus with an albedo A = 0.1. [18]

of radicals or their precursers can be stored in the nucleus
in this way, and these can then be brought out and deposited
in a region whose extent is determined only by the lifetimes
of the small grains stripped off from the nuclear matrix as
the gases evaporate. This may be an alternative way of explaining most of the radicals with the exception of OH, whose
very high abundance ($\gtrsim$ 85% by number of all the radicals)[19],
[20] suggest that it is a dissociation product of H_2O. (This
point will be further discussed in the following section.)

The Recent Ultraviolet Observations and Their Interpretation

Until about three years ago all observations of comets
were ground based, and consequently only those emissions
with wave lengths longwards of the ozone limit of about
3000 Å could be detected. Early in 1970, however, two long
period comets, Tago-Sato-Kosaka, and Bennett were observed in the ultraviolet by detectors on OAO-2 by Code and
Lillie. Strong Ly-α emission was seen in the heads of both
comets. Comet Tago-Sato-Kosak at a heliocentric distance
of about 0. 8 A. U. had a Ly-α emission region comparable to
the size of the sun; while comet Bennett (which was also observed by the Paris group of Blamont with the U-V detectors
on OGO-5) showed an emission region about 10 times as
large at the same heliocentric distance (see Fig. 4).

An early attempt at explaining the Ly-α emission invoked charge exchange excitation of solar wind protons with
cometary gasses [21]; the idea here being that the neutral H
atoms formed by this process will find themselves in excited
states and would cascade down to the ground state, emitting
Ly-α photons in the process. This treatment, however, fails
to take into account the likely existence of a Venus-type magnetosheath around the comet's head, as is indicated by the
flow of tail ions which seem to originate from a restricted
region in the inner coma.

The interplanetary magnetic field, which is convected
by the solar wind cannot diffuse through the cometary ionosphere in a time comparable to the time it takes to sweep
past it. Consequently the interplanetary field piles up

against the ionosphere till the pressure it exerts balances the ram pressure of the solar wind. The solar wind protons cannot penetrate this magnetosheath (which is estimated to be around 10^4 - 10^5 km from the center) and consequently charge exchange can take place only in an outer shell surrounding the cometary nucleus. The Ly-α emission in such a case would exhibit a projected structure showing a strong depletion toward the center, as in a planetary nebula, whereas observations indicate that the emission is strongest towards the nucleus.

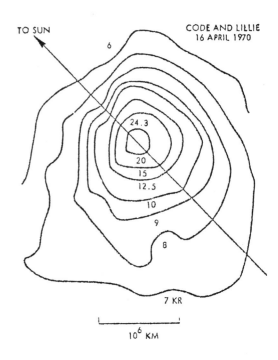

Fig. 4 Isophotes of Comet Bennett.

Consequently we[22] developed a different model where the source of Ly-α is neutral hydrogen produced by the photodissociation of water flowing out of a central icy nucleus. The neutral hydrogen so produced being ultimately removed by photodissociation and charge exchange with the solar wind protons in the outer coma. This is not a new idea--the existence of a large hydrogen coma produced by photodissociation of some hydrogenic molecule (very probably water) present in the head was anticipated by Biermann and Trefftz in 1964[23], and a rough estimate of its extent had been given subsequently by Biermann on the basis of the expected flow-velocities and lifetimes against ionization.

Although any hydrogenic molecule like NH_3 or CH_4, believed to exist in the nucleus could ultimately be the source of hydrogen, NH_3 itself is a rather doubtful candidate--the reason being that it should be observable via an emission band around 3240Å if present, but has not yet been observed in any comet. On the other hand CH_4, if it exists at all, probably does so in the form of the hydrate $CH_4 \cdot 6H_2O$ as pointed out earlier.

Also, besides Ly-α, the most prominent feature observed by OAO-2 in comets T-S-K and Bennett was the ground rotation-vibration band of OH around 3090Å, and the inferred abundances of the two species agree to within a factor of 2, strongly suggesting the same precurser.

We computed a complete hydrodynamic model of a cometary atmosphere composed of H_2O and its daughter products OH, H and O coupled through frictional interactions as well as production and loss processes and this explained rather well the observed Ly-α brightness distribution. All the processes taken into account as well as their corresponding rate coefficients are shown in Table 1, and the computed brightness distribution for Comet Bennett (at a heliocentric distance of about 0. 8 A. U.) is shown in Figure 5.

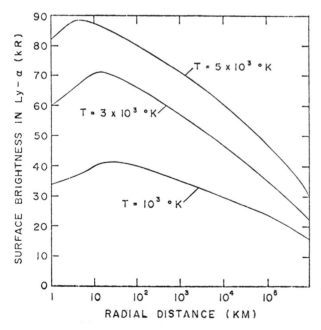

The above model assumed that the H_2O was evaporating from a monolithic central nucleus. It has, however, been shown experimentally[18] that in a vacuum simulating

Fig. 5 The Ly-α emission profile in the cometary coma.

cometary conditions, grains of varying sizes would be continuously stripped from the main body of the clathrate hydrate snows by gases evaporating from the nucleus to build
up an extensive halo of icy grains within the inner coma.
These grains then would themselves provide an important
supplementary source for the production of coma gases by
evaporation. Consequently we[25] have extended the earlier
model to include the case of a central source complemented

Table 1

Collision frequencies, production rates and loss coefficients.
R measures the heliocentric distance of the comet in AU.
τ_1, τ_2 and τ_3 are optical depths appropriate to photodissociation of H_2O, photodissociation of OH and photoionization of H and O. X, X_1, and X_2 represent any one of the
heavy molecules H_2O, OH and O. $\beta_2(H)$ and $\beta_2(O)$ are
only considered in r > 10^4 km.

Collision frequencies s^{-1} for H and X (representing H_2O, OH or O)

$$\nu(H, X) = 6.3 \times 10^{-16}(2.1 \times 10^8 T(H) + [u(H) - u(X)]^2)^{1/2} \cdot n(X)$$

$$\nu(X, H) = 3.5 \times 10^{-17}(2.1 \times 10^8 T(H) + [u(H) - u(X)]^2)^{1/2} \cdot n(H)$$

$$\nu(X_1, X_2) = 3.3 \times 10^{-16}(2.4 \times 10^7 T(X_1) + [u(X_1) - u(X_2)]^2)^{1/2} \cdot n(X_2)$$

Production rates $(cm^{-3} s^{-1})$ Loss coefficients (s^{-1})	Process		
$\beta(H_2O) = 1.6 \times 10^{-5}R^{-2}e^{-\tau_1}$	H_2O + hν	→ H	+ OH
$q(OH) = 1.6 \times 10^{-5}R^{-2}e^{-\tau_1} n(H_2O)$	H_2O + hν	→ H	+ OH
$\beta(OH) = 1.4 \times 10^{-6}R^{-2}e^{-\tau_2}$	OH + hν	→ O	+ H
$q(O) = 1.4 \times 10^{-6}R^{-2}e^{-\tau_2} n(OH)$	OH + hν	→ O	+ H
$\beta_1(O) = 5.0 \times 10^{-7}R^{-2}e^{-\tau_3}$	O + hν	→ O^+	+ e
$\beta_2(O) = 4.2 \times 10^{-7}R^{-2}$	O + H^+_{sw}	→ O^+	+ H_{sw}
$q_1(H) = 1.6 \times 10^{-5}R^{-2}e^{-\tau_2}n(H_2O)$	H_2O + hν	→ H	+ OH
$q_2(H) = 1.4 \times 10^{-6}R^{-2}e^{-\tau_2}n(OH)$	OH + hν	→ H	+ O
$\beta_1(H) = 2.0 \times 10^{-7}R^{-2}e^{-\tau_3}$	H + hν	→ H^+	+ e
$\beta_2(H) = 4.0 \times 10^{-7}R^{-2}$	H + H^+_{sw}	→ H^+	+ H_{sw}

by a distributed source in the inner coma. We have in effect considered two limiting cases of this distributed source model: (a) a stationary model where the distributed source is at rest with respect to the nucleus, and (b) a streaming model where the distribued source is expanding radially with a steady terminal velocity. The applicability of these two situations will be discussed later.

The major drawback of the following calculation is the same as in the central source model, viz, the substitution of a polytropic equation for the proper energy conservation equation, which should not only include the effects of heating the atmosphere by solar radiation but also energy transfer among the several constituents. However, it is possible to make rather judicious choices of the various polytropic indices based both on physical considerations as well as observation.

At the photodissociation of the H_2O most of the excess energy that does not go into excitation of OH is carried away by the much lighter H, and is in excess of 2 eV. (This point will be discussed in greater detail later.) Despite the inefficiency of the H in energy transfer during collisions with the heavier species, the number of collisions in $r \lesssim 500$ km is about 50. Consequently we expect all the species to be highly thermalized and a more or less isothermal expansion in this region. For $r > 500$ km the collision frequency of the H with the heavier species falls off rapidly and the expansion of the heavier species would quickly approach adiabaticity. On the other hand the H now being unable to get rid of much of its excess energy by collisions with the heavier species would heat up despite the expansion. Observations[1] indicated an upper limit for $<T>_H$ of about 1600° K for comet Tago-Sato-Kosaka when its heliocentri distance was $\lesssim 1$ A. U., and the best fit for the brightness profile of comet Bennett at a heliocentric distance of 0. 8 A. U. was obtained using $<T>_H \approx 100^\circ K$[22]. Such a temperature is best simulated by letting α_H (the polytropic index for H) decrease by about 20% in $5 \times 10^2 \lesssim r \lesssim 10^5$ km.

As regards the grains flowing out of the nucleus under the effects of the gas pressure and the gravitation of the nucleus there is a maximum size for the grains which can leave the nucleus. This value is given by

$$R_c = \frac{3 Q_o\left(H_2O\right) U_o\left(H_2O\right)}{4 \rho_g GM},$$

where $Q_o\left(H_2O\right)$ is the total rate of sublimation of H_2O from the nucleus. For comet Bennett at a heliocentric distance ≈ 0.8 A. U., $R_c \approx 2$ cm (assuming $\rho_g \approx 0.5$ g/cm^3 and $M \approx 10^{18}$ g). Velocity profiles corresponding to a set of different grain radii are shown in Fig. 6. It is seen that the grains are rapidly accelerated in the first 50 km and the terminal velocity is reached within a distance of about 200 km, beyond which the grains are effectively decoupled from the gas as a result of the radial divergence. It is seen that while grains of radius 1 μ can attain a terminal speed of about 0.5 km sec^{-1} a grain of radius 1 cm can attain a terminal speed of only about 5 m sec^{-1}.

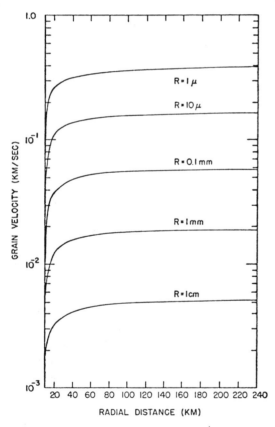

Fig 6   Velocity profiles of grains of different sizes.

While laboratory simulations[24] have established a sharp peak in the observed size distribution of the grains between 0.1 and 1 mm, this may not be directly applicable to the cometary situation we are considering. Indeed, it seems to us on rather general grounds that the distribution is likely to be peaked around the critical radius because in the continuous development of a cometary coma as the comet approaches the sun smaller grains would be stripped off earlier when they correspond to the critical size appropriate to a smaller flux. We have therefore assumed a grain radius of 1 cm with the corresponding velocity of about 5 m sec^{-1}.

Assuming that the production in distributed source is 30% of the total, the computed velocity and density profiles are shown in Fig. 7. While the velocity profile of H in Fig. 7

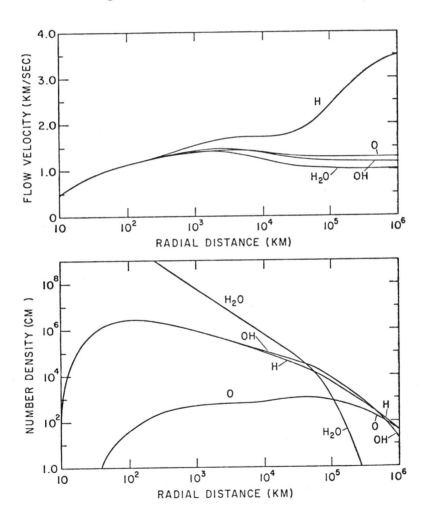

Fig. 7 Velocity and density profiles of the cometary atmosphere in the streaming model.

corresponds to the isothermal case (i.e. α_H = 1 all throughout) the variation of this profile for different values of α_H
are shown in Fig. 8. It should be noted that the velocity profiles of H_2O, OH and O are virtually unaffected by that of H.
While H attains a maximum speed of around 3.5 km/sec at a
distance of 10^6 km from the nucleus in the isothermal case,
it can attain a speed of about 8 km/sec at the same distance
in the case where α_H = 0.8. Indeed α_H = 0.8 seems to best
simulate the observed average temperature of the H (as discussed in section 2) and the corresponding speed of about
8 km/sec is completely consistent with the value deduced
from the observed distortion of the outer iosphotes of comets
Tago-Sato-Kosaka and Bennett due to Ly-α radiation pressure.[26]

Unlike in the central source model the hydrodynamic
motion is somewhat damped by the outgassing from the distributed source which loads the flow. Coupling between the
H_2O and H throttles the flow of the latter up to a distance of
about 5 x 10^4 km, beyond which its velocity picks up quite
strongly. The density profile is not too different from that
of the central source model, the main difference being a

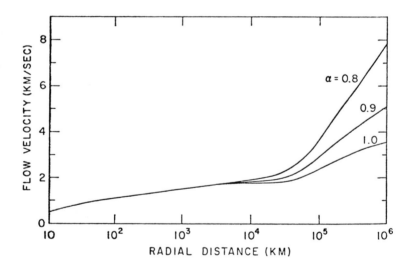

Fig. 8   Velocity profiles of H in the cometary atmosphere
corresponding to different rates of heat input (i.e.
different α's) in the streaming model.

slower buildup of OH and H to somewhat smaller maxima (by
a factor of 2) at significantly larger distances from the nucleus.

Another source distribution of considerable interest
is a cometary nucleus surrounded by a cluster of rather large
"grains" with a negligibly small radial velocity. While
Delsemme and Miller[27] have shown that meter size chunks
of the nucleus can be stripped off by the large flux of evaporating gases when the comet is sufficiently close to the sun
(d $\lesssim$ 0.3 A.U.), the possible existence of such a structure is
also suggested by the observations of P/Honda-Mrkos-
Padjusakova.[28]   Such a structure is also appropriate to the
model used in connection with formation of short period comets from meteor streams via Jupiter's gravitational perturbations.[29]

The velocity and density profiles in this case are
shown in Fig. 9, and the variation of the hydrogen velocity
profile for different α_H is shown in Fig. 10. While the
damping of the flow is observed in this case too, the effect is
more marked than in the streaming model, with a distinct
bottleneck in the H profile around 10^6 km, which distance
however is about the same in both cases.

The quantitative aspects of the velocity profiles
would naturally vary from comet to comet, and also, in the
case of a given comet, with heliocentric distance. Qualitatively, however, the above velocity profiles should be typical
of all comets with distributed sources; the feature distin-
quishing them from the central nucleus model being the
strong damping of the flow in the inner region. Also the
damping is stronger in a stationary model where a distinct
bottleneck is observed in the H profile.

Conclusions

The qualitative difference between the velocity profiles in the central source and distributed source models is
of particular interest because these could provide us with an
indirect way of obtaining information about the structure of
the all important nuclear region.

The internal velocity profiles of the various species cannot be measured directly. However, the so-called "Greenstein effect" (which observationally constitutes the variation of the intensity of some rotational line of an absorption band in a direction normal to the dispersion[30] provides us a method, at least in principle, of constructing the internal velocity profile of the species responsible for the absorption. If one could construct sufficiently accurate intensity profiles of individual rotational lines normal to the dispersion

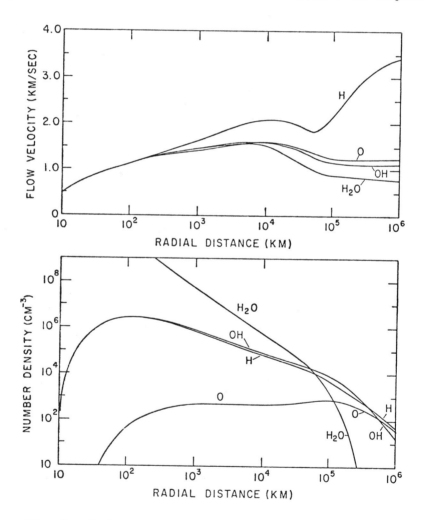

Fig. 9   Velocity and density profiles of the cometary in the stationary model.

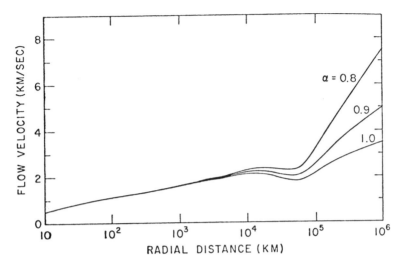

Fig. 10   Velocity profiles of H in the cometary
atmosphere corresponding to different
rates of heat input (i. e. different α's)
in the stationary model.

in high dispersion spectra one could use them to discrimi-
nate between different atmospheric models having self consistent density and velocity profiles. One requires, however,
a considerably greater accuracy in the measurements than is
available at present because, while the radial component of
the orbital velocity of a comet is typically a few tens of km
sec^{-1}, the dispersion of the internal velocity profile is typically of the order of a few tenths of a km sec^{-1}.

In conclusion a few remarks about the velocity of the
neutral hydrogen are in order. The observations discussed
earlier, as well as the arguments based on the observed
temperature, used in the foregoing analysis, all seem to suggest a velocity for H of about 8 km/sec. The possible photodissociation paths of H_2O by the solar ultraviolet are shown
in Table 2. Dissociation is possible by the Ly-α and Ly-β
line radiation too, but this is negligible ($\lesssim 10\%$). Most of the
dissociation (over 80%) is in the first continuum (1400 Å $\leq$ λ
$\leq$ 1860 Å) via a predissociation state into OH and H in their
ground states (reaction 1) although the bond energy of H_2O
(5. 1 eV) corresponds to λ_b = 2420 Å . Most of the excess

Table 2   Photodissociation of water
by the solar ultraviolet

A. In the first continuum (1800-1400 A)

(1) $H_2O + hv \rightarrow H\left(^2S\right) + OH\left(X^2\pi\right)$

(2) $H_2O + hv \rightarrow H_2 + O\left(^1D\right)$

B. In the second continuum (1400-1150 A)

(3) $H_2O + hv \rightarrow H\left(^2S\right) + OH\left(A^2\Sigma^+\right)$

(4) $H_2O + hv \rightarrow 2H\left(^2S\right) + O\left(^3P\right)$

(5) $H_2O + hv \rightarrow 2H\left(^2S\right) + O\left(^1D\right)$

energy is believed to go into the translational mode of H,
rather than into the excitation of the rotation-vibration modes
of OH[31]. Using the fact that the photodissociation cross-
section of H_2O is a maximum at about 1670 Å, Keller[26] estimated the average energy input into H to be about 2.5 eV
per dissociation, whereas, if account is taken of the distribution of solar energy in this region the input is closer to
2 eV[32] which corresponds to about 20 km/sec. This apparent discrepancy has led to arguments against a purely
H_2O source for H[5]. It must, however, be pointed out that
while H, produced by the photodissociation of H_2O, can lose
a substantial portion of its energy by collision in the inner
region ($r \lesssim 10^4$ km), in the outer regions ($r \gtrsim 5 \times 10^4$ km),
H is produced mostly by the photodissociation of OH (see
Figs. 7 and 9). The variation of the photodissociation cross-
section of OH with frequency is not available at present, in
order to calculate the average input of energy to the H during
the photodissociation of OH. Should this, however, be around
0.3 eV, then the observed velocity of H in the outer regions
may be explained without invoking another major source for
H besides H_2O.

References

[1] Code, A. D. and Savage, D., "Orbiting Astronomical Observatory: Review of Scientific Results", Science, Vol. 177, 1972, pp. 213-218.
Bertaux, J. L., Blamont, J. E., and Festau, M., "Interpretation of Hydrogen Ly-α Observations of Comets Bennett and Encke", preprint 7-A-73, Service D'Aéronomie, CNRS, Verrieres-Le-Buissow, 1973.

[2] Kleinmann, D. E., Lee, T., Low, F. J., and O'Dell, D. R., "Infrared Observations of Comets 1969g and 1969i", Ap. J., Vol. 165, 1971, pp. 633-636.
Westphal, J. A., "Infrared Observations of Comets Ikeya-Seki (1965f) and Bennett (1969i)", Proceedings of the Tucson Comet Conference (Ed. G. P. Kuiper and E. Roemer), 1972, pp. 23-31.

[3] Goldberg, R. A. and Aikin, A. C., "Comet Encke: Meteor Metallic Ion Identification by Mass Spectrometer", Science, Vol. 180, 1973, pp. 294-296.
Narcisi, R. S., "Processes Associated with Metal-Ion Layers in the E Region of the Ionosphere", Space Res., Vol. 8, 1968, pp. 360-365.

[4] Richter, N. B., The Nature of Comets, Methuen, London, 1963.

[5] Biermann, L., "Comets and their Interaction with the Solar Wind", Q. Jl. R. Astr. Soc., Vol. 12, 1971, pp. 417-432.

[6] Biermann, L., Brosowski, B., and Schmidt, H. U., "The Interaction of the Solar Wind with a Comet", Solar Phys., Vol. 1, 1967, pp. 254-284.

[7] Wallis, M., "Shock-Free Solutions of Solar Wind—Cometary Plasma Flows?", Trans. Royal Institute of Technology, Stockholm, No. TRITA-EPP-71-33, 1971.

[8] Wallis, M., "Weakly-Shocked Flows of the Solar Wind Plasma through Atmospheres of Comets and Planets", Astron. Astrophys, Vol. 29, 1973, pp. 29-44.

[9] Mentzoni, M. H. and Donohoe, J., "Electron Recombination and Diffusion in CO at Elevated Temperature", Can J. Phys., Vol. 47, 1969, pp. 1789-1791.

[10] Axford, W. I., private communication, 1972.

[11] Finson, M. L. and Probstein, R. F., "A Theory of Dust Comets 1. Model and Equations", Astrophys. J., Vol. 154, 1968, pp. 327-352.

[12] Roemer, E., "The Dimensions of Cometary Nuclei", Nature et Origine des Cometes, Liege, 1966, pp. 23-28.

[13] Lyttleton, R. A., The Comets and their Origin, Cambridge, 1952.

[14] Whipple, F. L., "On the Icy Conglomorate Model for Comets", La Physique des Cometes, Liege, 1953, pp. 281-286.

[15] Larson, S. M. and Minton, R. B., "Photographic Observations of Comet Bennett, 1970 II", Proceedings of the Tucson Comet Conference (Ed. G. P. Kuiper and E. Roemer), 1972, pp. 183-208.

[16] Delsemme, A. H. and Swings, P., "Hydrates de Gaz Dams Lesnovaux Cometaires et les Grains Interstellaires", Ann. Astrophys., Vol. 15, 1952, pp. 1-15.

[17] Miller, S. L., "The Occurrence of Gas Hydrates in the Solar System", Proc. Nat. Acad. Sci., Vol. 47, 1961, pp. 1798-1806.

[18] Delsemme, A. H. and Wenger, A., "Physico-Chemical Phenomena in Comets--1", Planet. Space Sci., Vol. 18, 1970, pp. 709-715.

[19] Code, A. D., "Comments on OAO Observations of Comet 1969g and 1969i", preprint (unpublished) 1970.

[20] Arpigny, C., "Spectra of Comets and their Interpretation", Ann. Rev. of Astron. and Astrophys., Vol. 5, 1965, pp. 351-376.

[21] Tolk, N. H., White, C. W., and Graedel, T. E., "On the Ly-α Halos around Comets Tago-Sato-Kosaka and Bennett", paper read at the 132 meeting of the AAS in Colorado, 1970.

[22] Mendis, D. A., Holzer, T. E., and Axford, W. I., "Neutral Hydrogen in Cometary Comas", Astrophys. and Space Sci., Vol. 15, 1972, pp. 313-325.

[23] Biermann, L. and Trefftz, E., "Über die Mechanisman der Ionization and der Anregung in Kometenatmosphären", Z. Astrophys., Vol. 59, 1964, pp. 1-28.

[24] Delsemme, A. H. and Miller, D. C., "Physico-Chemical Phenomena in Comets - II", Planet. and Space Sci., Vol. 18, 1970, pp. 717-730.

[25] Ip, W-H. and Mendis, D. A., "Neutral Atmospheres of Comets: A Distributed Source Model", Astrophys. and Space Sci., Vol. 26, 1974, pp. 153-166.

[26] Keller, H. U., "Hydrogen as a Product of Dissocation in Comets", Mitt. Astron. Gesellschaft, Vol. 30, 1971, pp. 143-146.

[27] Delsemme, A. H. and Miller, D. C., "Physico-Chemical Phenomena in Comets - III", Planet. Space Sci., Vol. 19, 1971, pp. 1229-1257.

[28] Mrkos, A., "Observation and Feature Variations of Comet 1969e before and during Perihelion Passage", Proc. 21st Nobel Symposium, Saltsjöbaden, Sept. 1971, pp. 261-272.

[29] Trulsen, J., "Formation of Comets in Meteor Streams", Motion, Orbit Evolution and Origin of Comets, Proc. IAU Symposium No. 45, Leningrad, 1970, pp. 487-490.
Alfvén, H., "Apples in a Spacecraft", Science, Vol. 173, 1971, pp. 522-525.

Mendis, D. A., "The Comet-Meteor Stream Complex", Astrophys. and Space Sci., Vol. 20, 1973, pp. 165-176.

[30]Greenstein, J. L., "High Resolution Spectra of Comet Mrkos (1957d)", Ap. J., Vol. 128, 1958, pp. 106-116.

[31]Welge, K. H. and Stuhl, F., "Energy Distribution in the Photodissociation of $H_2O \rightarrow H(1^2S) + OH(X^2\pi)$", J. Chem. Phys., Vol. 46, 1966, pp. 2440-2441.

[32]Mendis, D. A. and Ip, W-H., "The Neutral Atmospheres of Comets", Astrophys. and Space Sci., Vol. 39, 1976, pp. 335-385.

[33]Ananthakrishnan, S., Bhandari, S. M., and Pramesh Rao, A., "Occultation of Radio Source PKS 2025-15 by Comet Kohoutek (1973f)", Astrophys. and Space Sci., Vol. 37, 1975, pp. 275-282.
Hyder, C. L., Brandt, J. C., and Roosen, R. J., "Tail Structures Far from the Head of Comet Kohoutek. 1.", Icarus, Vol. 23, 1974, pp. 601-610.
Ip, W-H. and Mendis, D. A., "On the Interpretation of the Observed Cometary Scintillations", Astrophys. and Space Sci., Vol. 35, 1975, pp. L1-L4.
Ip, W-H. and Mendis, D. A., "The Cometary Magnetic Field and its Associated Electric Currents", Icarus, Vol. 26, 1975, pp. 457-461.

[34]Ip, W-H. and Mendis, D. A., "The Generation of Magnetic Fields and Electric Currents in Cometary Plasma Tails", Icarus, 1976 (in press).
Ip, W-H. and Mendis, D. A., "The Structure of Cometary Ionospheres 1. H_2O Dominated Comets", Icarus, 1976 (in press).

[35]Wurm, K., "The Physics of Comets", Moon Meteorites and Comets (Ed. B. M. Middlehurst and G. P. Kuiper), University of Chicago Press, 1963, pp. 573-617.

Addendum (added in proof)

A very important contribution to our knowledge of neutral atmospheres has recently come from radio observations [32]. This is the spectroscopic detection, for the first time of two stable neutral molecules CH_3CN and HCN in Comet Kohoutek (1973f). While OH and CH were earlier observed only in the optical, they have now also been observed in the radio via their hyperfine splitting of the ground state Λ doublet in the same comet. A more tentative detection of H_2O at 1.35 cm, due to a ground state rotation-vibration transition, has been reported in the subsequent Comet Bradfield (1974b) [32]. If substantiated, this would be the first direct observation of what has long been regarded as the most abundant parent molecule in most, if not all, comets. Further support for this view has been provided by the identification of several H_2O^+ lines, first in the optical spectrum of Kohoutek (1973f) and subsequently in Bradfield (1974b)[32]. Ultraviolet observations of Kohoutek (1973f) while confirming the observations of atomic hydrogen and atomic oxygen in earlier comets have further identified atomic carbon [32].

Observations of cometary scintillations as well as propagating helical structures in Comet Kohoutek (1973f) have suggested the existence of substantial magnetic fields (100-1000 γ) [33]. A possible mechanism for the generation of these magnetic fields has been proposed and the role of the associated electric currents in maintaining the ionospheric structure has been evaluated [34]. It appears that the energetic electrons (1-10 keV) constituting the current ($\sim 10^8$A) may very well be the long sought "internal ionization source" proposed to explain the ionization features within 10^3 km from the nucleus [35].

INTERPRETATION OF LYMAN ALPHA OBSERVATIONS
OF COMET BENNETT (1970 II)

H. U. Keller*

Laboratory for Atmospheric and Space Physics,
University of Colorado, Boulder, Colorado

Abstract

The satellite UV observations of comet Bennett (1970 II) are briefly reviewed, discussing their results and interpretations. The observations by the University of Colorado photometer are described and a preliminary interpretation is given. The curvature of the extended hydrogen tail yields the solar Ly-α (1216Å) flux from calculations similar to those used in the analysis of dust tails. The hydrogen cloud is influenced by atomic resonance scattering in analogy to the dust particles driven away from the sun by scattering of the solar continuum. Since only the positions of the intensity maxima along the observational tracks are used and not the absolute values, this determination of the solar Ly-α is independent of instrumental calibration.

In 1970, several comets were observed for the first time in the ultraviolet wavelength range: comet Bennett (1970 II)[1,2,3,4,5], comet Tago-Sato-Kosaka (1969 IX) (hereafter TSK)[3], and the periodic comet Encke (1970 L)[5]. The most significant result was the detection of an extended hydrogen atmosphere of all three comets, predicted by Biermann 1968[6].

Presented at the AIAA/AGU Space Science Conference on the Exploration of the Outer Solar System, Denver, Colorado, July 10-12, 1973 (not preprinted). Performed under NASA Grant Number NGL-06-003-052.
*Research Associate; on leave from Max-Planck-Institut for Physics and Astrophysics, Munich, Germany.

Comet Bennett was the brightest of the observed comets with
a reduced visual brightness of 3.5m[7] and was also the best
observed. The first Ly-α (1216Å) observation was made just
before the comet's perihelion passage on March 20, 1970 by
the University of Colorado ultraviolet photometer on board
the Orbiting Geophysical Observatory OGO-5. This paper will
deal mainly with a preliminary interpretation of these only
recently reduced observational data, leading to new values for
the solar Ly-α flux[8]. In April 1970, comet Bennett was observed by another photometer on board OGO-5, that of the
University of Paris[4]. These data have been published[5]. In
1971, Keller[9] proposed the use of Haser's fountain model[10] for
interpretation of Ly-α isophote maps made from the French OGO-
5 observations. Bertaux et al.[5] used only the intensity profile along the sun-comet line, whereas Keller[9,11] used computed isophote models to determine the outflow velocity, v_H,
the lifetime, t_H, against ionization, and the production rate,
Q_H, of the cometary hydrogen atoms. Figure 1 shows the observed isophotes of April 1, 1970 together with the best
matching computed model. Assuming a solar Ly-α flux of 3.2 x

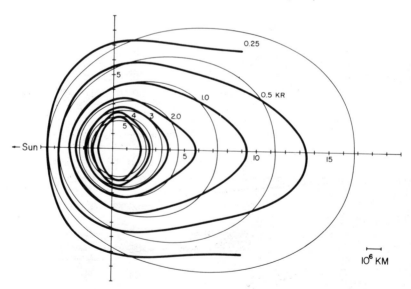

Ly α Isophotes Comet Bennett (1970 II) April 1, 1970

Fig. 1 Ly-α isophotes of comet Bennett (1970 II) on April 1,
1970. Sun-comet distance 0.61 a.u. ▬ Observed
isophotes Bertaux et al.[5] - computed isophotes
Keller[11].

10^{11} ph cm^{-2} sec^{-1} Å^{-1} at 1 a.u., the following results were
found for comet Bennett in April 1970, when the comet's heliocentric distance increased from 0.6 to 0.86 a.u.: v_H = 8.2km
s^{-1}; t_H = 2.2 x 10^6 s at 1 a.u.; $Q_H \sim 10^{30}$ H atoms s^{-1}
(Keller[11]). These results are in good overall agreement with
those of Bertaux et al.[5], who also found a variation of t_H
with heliographic latitude, that differs from R^{-2} (R is the
heliocentric distance of comet Bennett). Keller[11], found that
the production rate decreased proportionally to R$^{-1.5}$.

The central part (3° in diameter, outermost isophote ~ 8
kR) of comet Bennett was also observed by the Orbiting Astronomical Observatory, OAO-2, (Code et al.[3]). The Ly-α isophotes of the inner part of the hydrogen atmosphere in Ly-α
were interpreted by a more complex model taking multiple
scattering into account (Keller[12,13]), giving good agreement
with the interpretations of the French OGO-5 observations.
The OAO-2 observations of comets Bennett and TSK have been
only partially published. Delsemme[14] investigated the production rate variation of hydrogen and OH with heliocentric
distance for TSK. Both R-exponents are approximately equal
but considerably larger (-2.8) than that determined for the
hydrogen production of comet Bennett from the French observations. Recent investigations of OAO-2 data for comet
Bennett by Keller and Lillie[15] using Haser's[16] parent-daughter-
molecule model yielded a mutual exponent of -2.3 for the production rate of hydrogen and hydroxyl (OH) in the heliocentric
distance interval $0.76 \leqslant R \leqslant 1.26$ in April and May 1970. The
important scale length (outflow velocity x lifetime) of OH
could be determined to 2(-1.0 + 0.5) x 10^5 km at 1 a.u. for
the first time. The data show appreciably less scatter than
the French OGO-5 observations, indicating that the hydrogen
production rate after perihelion decreased faster for comet
TSK (Delsemme[14]) than for Bennett. Both comets differed
greatly in their dust production. On the other hand, the
visible coma brightness of Bennett decreased more rapidly
than that of TSK[7]. This behavior deserves a more detailed
investigation.

These results strongly favor the icy conglomerate nucleus
model of comets (Whipple[17]). But the question whether water
ice is the most abundant compound has not yet been solved
unambiguously.

The French observations of comet Bennett were made in
April 1970 during a special spin-up mode of the OGO-5 satellite: in March 1970, comet Bennett passed in front of the
field of view (FOV) track of the University of Colorado UV

photometer airglow experiment on OGO-5. In the earth stabil-
ized mode of the spacecraft, the FOV pointed outward along
the earth-spacecraft radius vector. The A-channel FOV, sensitive to the wavelength region 1150-1800Å, was 3°. The B-
channel sensitive to 1225A-1800Å emission and also 3° FOV,
showed no cometary emission above the instrument threshold.
Figure 2 shows the FOV and the path of the comet from March 15
to 25 when the five observations were recorded. The track

separation in a co-
ordinate system cen-
tered on the comet
(Fig. 3) is due to
the motion of the
comet during one or-
bital period (about
60 hours) of the
OGO-5 satellite.
The OGO-5 orbital
plane was fixed in
inertial space dur-
ing this period.
The scans were all
nearly perpendic-
ular to the tail
axis and the lines
of sight of the
intensity maxima
(indicated as X in
Fig. 3) were within
10° of the normal
of the comet's or-
bital plane. Since
the geometry of the

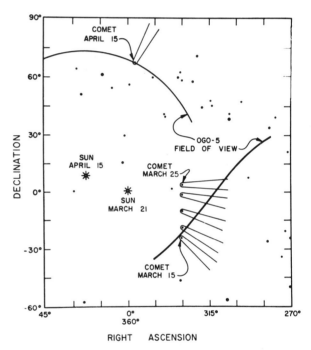

comet with respect   Fig. 2 Positions of comet Bennett, the
to the earth varied   Sun and the field of view track
only slightly be-   of the University of Colorado
tween orbits, the   photometer on OGO-5. The cir-
synthesized pre-   cular track occurred during a
liminary isophotes   special spin-up maneuver on
in Fig. 3 represent   April 15, 1970. The dots rep-
the intensity dis-   resent ultraviolet stars.
tribution of the

hydrogen atmosphere in Ly-α. Measurements were made along the
tracks shown from March 15 until March 25, 1970. The maximum
intensity of 77R along the March 25 track lies 16° from the
nucleus, corresponding to 3 x 10^7 km (1/5 a.u.). This shows
the enormous extent of the cometary hydrogen atmosphere.
Parameters for the hydrogen atmosphere, similar to those

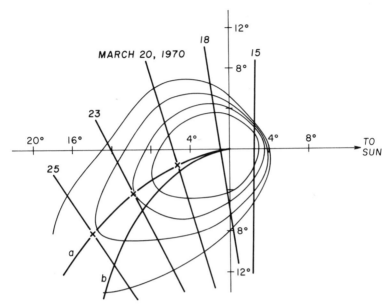

Fig. 3 The observing tracks in a cometary coordinate system
        during the time interval from March 15 to March 25,
        1970. Comet-earth distance is about 0.72 a.u. The
        crosses (X) indicate the measured intensity maxima.
        Curves (a) and (b): syndynames due to Ly-α fluxes of
        9.5 and 5 x 10^{11} ph cm^{-2} s^{-1} $Å^{-1}$ respectively. Pre-
        liminary isophotes are indicated.

derived from the French observations, can be determined from
these results but the analysis is more complicated since
orbital acceleration must be taken into account. This work is
in progress and will not be discussed here. Additional parameters, such as the solar Ly-α flux, can be determined. The
hydrogen atmosphere is strongly elongated in antisolar direction but shows (Fig. 3) a distinct deviation from the line
comet-sun. This deviation is caused by the orbital movement
of the comet and the repellent solar radiation pressure force,
resulting in a curvature of the hydrogen atmosphere similar to
the curvature of the dust tail.

    Since the end of the last century the trajectories of
dust particles leaving the cometary nucleus have been computed. The grains are driven from the nucleus by a reduced
attractive or even repellent central force of the sun, de-
pending only on the ratio of the radiative to the gravitational force. Finson and Probstein[18] recently improved the com-
putational methods and the theory. The radiation pressure

force of the solar continuum influences the small dust particles forming the cometary dust tail; the influence on the
cometary nucleus is negligible. A computation similar to the
dust tail calculations is used for the hydrogen atoms produced by photodissociation of molecules, such as water in the
vicinity of the nucleus. The scale lengths of possible parent molecules of hydrogen are about 10^5 km (e.g., water[11]) at
1 a.u. but in any case smaller than 10^6 km. The production
zone of hydrogen for comet Bennett at perihelion (~ 0.5 a.u.)
had a radius of about 0.5 to 2 x 10^5 km (the scale lengths
scale proportional to $R^2 = 0.25$). Even if the parent molecule
evaporation from the nucleus is not isotropic the somewhat
extended source for the hydrogen atoms should be. We know
from visible observations of other daughter molecules (such as
CN, C_2) that spherical symmetry is a good first order assumption. The overall extent of the hydrogen source is any-
way so small compared to the extension of the hydrogen atmosphere ($> 10^7$ km, two orders of magnitude) that a point source
is assumed in the following calculations. Collisions do not
occur outside the source region because of the low densities.
Unlike the dust grains, hydrogen atoms are all affected by an
equal force (in first order neglecting the solar Ly-α profile),
whereas the net force on the grains depends on their size
(varying several orders of magnitude), leading to the dispersion of the dust tail. This simplifies the calculation and
the physics involved. Radiation pressure leads to a radial
acceleration of b = g $\cdot$ hν_{1216}/m_H $\cdot$ c (g is the solar excitation rate per atom). The trajectory of an H atom depends
upon the ratio (traditionally called 1-μ) of this solar
radiation pressure force (caused by the resonance scattering
of the solar Ly-α line) to the gravitational force and upon the
initial velocity of the atom. The equation of motion for the
conic section trajectories (hyperbolas convex to the sun if
(1-μ) > 1) can be integrated using the techniques of the dust
tail model. All hydrogen atoms that left the nucleus at some
earlier time with zero ejection velocity and were influenced
by a certain (1-μ) ratio, form a curve in the plane of the
cometary orbit called the "syndyname." The curvature of the
syndyname is determined by the geometry of the cometary orbit
and 1-μ; the larger 1-μ is, the more closely the syndyname is
aligned with the antisolar direction. H atoms leaving the
nucleus with non-zero ejection velocities (typically 10 km
s^{-1}) form expanding spherical surfaces with center points on
the corresponding syndyname. Since the H atom production is
assumed isotropic and 1-μ is identical for all H atoms, the
hydrogen tail curvature is determined by the corresponding
syndyname representing the symmetry line, and, therefore, the
line of maximum column density normal to that line. The

syndyname connecting the intensity maxima (X in Fig. 3, curve
a) corresponds to a $1-\mu$ ratio of 2.84 using a solar Ly-α flux
at 1 a.u. = 9.5 x 10^{11} ph cm^{-2} s^{-1} Å^{-1}. A syndyname for a
Ly-α flux of 5 x 10^{11} ph cm^{-2} s^{-1} Å^{-1} ($1-\mu$ = 1.5) is also displayed for comparison (curve b). Possible physical effects
and observational errors that might detract from the significance of the results will now be discussed.

    1. The expanding spherical surfaces of the H atoms with
non-zero ejection velocities are slightly distorted by the
spatial force gradient as they become larger. This effect remains less than 10% in radius of the sphere. Due to a favorable geometrical situation, the symmetry with respect to the
syndyname is even less disturbed.

    2. The solar Ly-α line shows a reversal of about 30%
intensity decrease at line center[19]. Hence H atoms, with
different radial velocities with respect to the sun, v_R,
scatter different amounts of light due to the intensity variation of the Ly-α line profile. The repellent radiation force
during the lifetime of an H atom changes slightly. The syn-
dynames therefore represent a certain average of the solar
line profile intensity rather than the value at line center.
At the time of observation, the hydrogen atoms scatter different solar intensities according to their actual radial
velocity with respect to the sun. This might separate the
intensity maximum of the hydrogen atmosphere from the column
density maximum (syndyname). The velocities of the H atoms
reach values of $v_R \sim$ 30 km s^{-1} at 2 x 10^7 km from the cometary nucleus, even if the solar flux is only 5 x 10^{11} ph cm^{-2}
s^{-1} Å^{-1}. The corresponding wavelength shift coincides with
the wavelength of the solar Ly-α maximum of the short wavelength wing. An additional shift of the cometary emission
maximum along the observational track away from the column
density maximum should therefore not occur if the solar Ly-α
profile observed by Bruner and Rense[19] is correct.

    3. The lifetime of the cometary H atoms is determined
mainly by charge exchange with solar wind protons. The densities are so low that any screening effects are negligible.
Assuming the lower solar Ly-α flux of 5 x 10^{11} ph cm^{-2} s^{-1} Å^{-1}
it takes the H atoms t = 1.5 x 10^6 s or nearly three times
their average lifetime t_H at 0.54 a.u. (perihelion distance of
Bennett) to reach a distance of 3 x 10^7 km from the nucleus.
Only approximately 10% will survive ($\propto e^{-t/t_H}$); 90% will be
converted into slow protons by the solar wind, yielding the
equivalent amount of "hot" H atoms (former solar wind protons).
The density of these hot H atoms at 3 x 10^7 km is therefore

comparable to that of the original cometary H atoms, taking into account the dilution of the hot atom density due to their high velocities of about 400 km s^{-1} (factor of 10). The momenta of the charge exchanging particles remain unchanged. The intensity contribution to the cometary emission, however, is only a few percent because the solar Ly-α line intensity of the blue wing at 1.6Å from the line center is strongly decreased. (The Doppler shift corresponding to the mean solar wind velocity of 400 km s^{-1} is 1.6Å.)

The influence of the solar wind needs more detailed investigation, evaluating appropriate models and eventually also considering the shock front. The shock front dimensions[20] seem to be too small compared to 3 x 10^7 km to affect the results.

The finite lifetime of the hydrogen atoms disturbs the symmetry assumptions somewhat. H atoms at symmetric locations with respect to the syndyname had different trajectories with respect to the sun and therefore different life expectations. This influence is difficult to investigate without detailed calculations, but is expected to be only of minor importance.

4. The 3° FOV does not degrade the results since the intensity gradients of the cometary emission along the three observational tracks (March 20-25) are small.

5. Pointing errors of the photometer may seriously influence the results. An investigation of star observations and also of the maxima near the cometary nucleus on March 15 and 18 was made. Conservative estimates of this error are probably less than 1°.

Figure 4 displays the result of this analysis. The tail deviation angle, δ, is the angle between the anti-solar direction and the radius vector from the nucleus to the crossing point of a syndyname with an observational track. The syndyname and δ depend upon the solar Ly-α flux. The Ly-α fluxes (at 1 a.u.) for March 20, 23 and 25 were 9.8 x 10^{11}, 10.5 x 10^{11} and 9.2 x 10^{11} ph cm^{-2} s^{-1} Å^{-1}. The vertical error bars represent the estimated instrumental pointing error of $\pm$ 1° along the track line. The horizontal bars indicate the corresponding uncertainty in the solar Ly-α flux determination. From calibrated photometric observations from the OSO-5 satellite, a mean flux at the solar Ly-α line center of about 5.0 $\pm$ 0.5 x 10^{11} ph cm^{-2} s^{-1} Å^{-1} could be expected for the days before the data of observations. The time the H atoms needed to reach the maximum point on the observational track

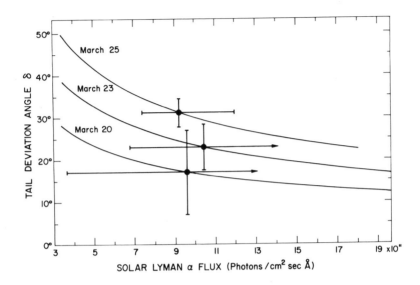

Fig. 4   The dependence of the tail deviation angle, δ, on the Ly-α flux for the three observation dates March 20, 23 and 25, 1970. Vertical error bars indicate the pointing error of the instrument. Horizontal error bars are the corresponding uncertainties in the flux determination.

on March 25, accelerated by a flux of 9.5×10^{11} ph cm^{-2} s^{-1} Å^{-1}, is about 10^6 s or 12 days. The appropriate Zurich sunspot number average was ~80 for the period March 11 to March 22 (different solar longitudes of the earth and the comet account for the three day difference). Using the high-resolution measurements of the profile of Bruner and Rense[19], a value of 6.3×10^{11} ph cm^{-2} s^{-1} A^{-1} would be expected for the average intensity of the blue wing of the self-reversed solar Ly-α line. The mean value found in this analysis, corresponding to an estimated flux at line center of 7.6×10^{11} ph cm^{-2} s^{-1} Å^{-1}, is approximately a factor of 1.5 higher than the OSO-5 results[21]. The results are in the high range of values deduced from satellite Ly-α airglow data. Meier and Mange[22] show these vary from experiment to experiment with values from 2.4×10^{11} up to 8.0×10^{11} ph cm^{-2} s^{-1} Å^{-1}. The present results, which are based purely upon dynamical considerations, support the suggestion of Meier and Mange that, because of sensitivity losses that often occur in orbit, the airglow observations should be viewed as lower limits.

The higher solar Ly-α flux at line center of around 7.6 x 10^{11} ph cm^{-2} s^{-1} Å^{-1} leads to different results in the interpretation of the French OGO-5 observations. The shape of the calculated isophotes would be only slightly changed and probably be still in the same overall agreement with the observed isophotes, but the mean outflow velocities and the production rate have to be changed according to the now larger solar radiation pressure and excitation ($v_H \propto \sqrt{flux}$, $Q \propto 1/\sqrt{flux}$). The outflow velocity would now be about 12.6 km s^{-1} compared to the old value of 8.2 km s^{-1} [11]. The value of 12.6 km s^{-1} can be more easily related to the photo-dissociation of H_2O and OH resulting in H atoms with high excess energies[9,11].

References

[1]Thomas, G. E., private communication, 1973.

[2]Code, H. D., Houck, T. E., and Lillie, C. F., "Comments on OAO Observations of Comet Tago-Sato-Kosaka (1969g)," IAU Circular No. 2201, 1970.

[3]Code, A. D., Houck, T. E., and Lillie, C. F., "The Scientific Results from the Orbiting Astronomical Observatory (OAO-2)," NASA SP-310, pp. 109-114.

[4]Bertaux, J. L. and Blamont, J. E., "Observation de l'emission d'hydrogene atomique de la Comete Bennett," Comptes Rendus de l'Academie des Sciences, Paris, Vol. 270, 1970, pp. 1581-1584.

[5]Bertaux, J. L., Blamont, J. E., and Festou, M., "Interpretation of Hydrogen Lyman-Alpha Observations of Comets Bennett and Encke," Astronomy and Astrophysics, Vol. 25, 1973, pp. 415-430.

[6]Biermann, L., "On the Emission of Atomic Hydrogen in Comets," Joint Institute for Laboratory Astrophysics Report No. 93, 1968.

[7]Beyer, M., "Physische Beobachtungen von Kometen. XVII," Astronomische Nachricten, Vol. 293, 1972, pp. 241-257.

[8]Keller, H. U. and Thomas, G. E., "Determination of the Solar Lyman Alpha Flux Independent of Calibration by UV Observations of Comet Bennett," Ap. J. Letters, Vol. 186, 1973, pp. L87-L90.

[9]Keller, H. U., "Wasserstoff als Dissoziationsprodukt in Kometen," Mitteilungen der Astronomischen Gesellschaft No. 30, 1971, pp. 143-148.

[10]Haser, L., "Calcul de distribution d'intensite relative dans une tete cometaire," Colloque l' Universite, Liege, 1965, pp. 233-241.

[11]Keller, H. U., "Hydrogen Production Rates of Comet Bennett (1969i) in the First Half of April 1970," Astronomy and Astrophysics, Vol. 27, 1973, pp. 51-57.

[12]Keller, H. U., "Die Lyman- Strahlung der Wasserstoff-atmospharen von Kometen - Ein Modell mit Mehrfachstreuung," Ph.D. Thesis, University of Munich, 1971.

[13]Keller, H. U., "Lyman- Radiation in the Hydrogen Atmospheres of Comets. A Model with Multiple Scattering," Astronomy and Astrophysics, Vol. 23, 1973, pp. 269-280.

[14]Delsemme, A. H., "The Brightness Law of Comets," Astrophysical Letters, Vol. 14, 1973, pp. 163-167.

[15]Keller, H. U. and Lillie, C. F., "The Scale Length of OH and the Production Rates of H and OH in Comet Bennett (1970 II)," Astronomy and Astrophysics, 1974, in press.

[16]Haser, L., "Distribution d'intensite dans la tete d'une comete," Bulletin de la Classe des Science, Academie Royale de Belgique, Vol. 43, 1957, pp. 740-750.

[17]Whipple, F., "A Comet Model I. The Acceleration of Comet Encke," Astrophysical Journal, Vol. 111, 1950, pp. 375-399.

[18]Finson, M. L. and Probstein, R. F., "A Theory of Dust Comets," Astrophysical Journal, Vol. 154, 1968, pp. 327-380.

[19]Bruner, E. C., Jr. and Rense, W. A., "Rocket Observations of Profiles of Solar Ultraviolet Emission Lines," Astrophysical Journal, Vol. 157, 1969, pp. 417-424.

[20]Brosowski, B. and Wegmann, R., "Numerische Behandlung eines Kometenmodells," Max-Planck-Institut fur Physik und Astrophysik, Munich, MPI/PAE-Astro 46, April 1972.

[21]Vidal-Madjar, A., Blamont, J. E., and Phissamay, B., "Solar Lyman Alpha Changes and Related Hydrogen Density Distribution at the Earth's Exobase (1969-1970)," Journal of Geophysical Research, Vol. 78, 1973, pp. 1115-1144.

[22]Meier, R. R. and Mange, P., "Spatial and Temporal Variations of the Lyman Alpha Airglow and Related Atomic Hydrogen Distributions," Planetary and Space Science, Vol. 21, 1973, pp. 309-327.

COMET EXPLORATION: SCIENTIFIC OBJECTIVES
AND MISSION STRATEGY FOR A RENDEZVOUS
WITH ENCKE

H. F. Meissinger[*] and E. W. Greenstadt[+]
TRW Defense & Space Systems Group, Redondo Beach, Calif.

W. I. Axford[‡]
Max-Planck-Institute f. Aeronomie, West Germany

and

G. W. Wetherill[§]
Carnegie Institute of Washington, Washington D.C.

Abstract

This paper reviews physical characteristics of a specific
cometary target, i.e., Encke, discusses scientific mission objectives and payload instruments, and describes an exploration
strategy tailored to these scientific objectives. Rendezvous
with the comet, as opposed to the brief encounter of a flyby
mission, permits systematic exploration of time-varying phenomena in the coma, tail, and the nucleus. A carefully designed
and executed exploration strategy will improve greatly the interpretation of observational data obtainable by ground-based

---

Presented as Paper 73-550 at the AIAA/AGU Space Science
Conference: Exploration of the Outer Solar System, Denver,
Colo., July 10-12, 1973. Results reported in this paper are
based on a study performed by TRW Defense & Space Systems Group
for Jet Propulsion Laboratory, Pasadena, Calif., under Contract
953247, sponsored by NASA Contract NAS7-100. The authors wish
to acknowledge the valuable guidance and critique of the ap-
proach presented here by K. L. Atkins and R. Newburn of Jet
Propulsion Laboratory, and the assistance in performing this
study by members of the TRW Technical Staff, especially Robert
Africano, Arthur Carlin, Daniel Goldin, and John Slattery.
Ruby Williams' help in report preparation deserves special ap-
preciation.
[*]Senior System Engineer.
[+]Member of Professional Staff.
[‡]Director.
[§]Director, Department of Terrestrial Magnetism

or Earth-orbiting telescopes. The rendezvous mission, which includes an extended stay of at least 80 days in the comet's vicinity with exploration maneuvers through the coma and tail and around the nucleus, is made feasible by the use of solar-electric propulsion. This new technology, which is now ready for flight application, offers its greatest advantage in missions with large total impulse requirements such as comet rendezvous where the target body's gravity is much too small to assist in spacecraft capture.

1.   Introduction and Summary

Missions to comets and asteroids, the small bodies that are believed to contain material representative of the primordial composition of the planets, will be an important step in determining the origin and formation processes of the solar system. Rendezvous with a comet, as opposed to the short encounter of a simple flyby mission, provides the opportunity for systematic exploration of the nucleus. In situ observation of cometary phenomena by a few well-selected missions with a carefully designed and executed exploration strategy will improve greatly the interpretation of observational data obtainable by ground-based and/or Earth-orbiting telescopes. These considerations provide a strong rationale for planning comet flyby and rendezvous missions in the near future, as discussed by NASA's 1971 Comet and Asteroid Mission Advisory Panel[1]. Comet exploration has been the subject of two NASA-sponsored working conferences in 1970 and 1971[2,3] and has been discussed widely in the literature in the past six to eight years[4-7].

This paper summarizes results of a recent study performed by TRW Systems[8] which examined physical characteristics of a specific cometary target, i.e., comet Encke, determined scientific mission objectives, identified payload instruments, and formulated an effective exploration strategy. The rendezvous mission, which includes an extended stay of 80 days or longer in the vicinity of the comet with exploration maneuvers through the coma and tail and circumnavigation of the nucleus, is made feasible by the use of solar-electric propulsion. This new technology, which is now ready for flight applications, offers its greatest advantage in missions with large total impulse requirements, such as comet and asteroid rendezvous, where the gravity of the target body is much too small to assist in spacecraft capture.

Considerations that lead to the selection of Encke as target for a first comet rendezvous mission, as well as planning, implementation, and timing aspects of this mission, are covered in a paper by Atkins and Moore[9]. Therefore, detailed discussion of these factors in trajectory selection and mission

definition can be omitted in this paper. The current plans
envision the rendezvous to take place during the 1984 appari-
tion of the comet.

Comet Encke is well suited as a target for a deep-space
probe because its short orbital period of 3.3 years has allowed
it to be observed on many perihelion passes since its discovery,
and its orbital parameters as well as perturbative influences
are better established than those of most other comets. For
purposes of this discussion, we assume that an Encke flyby mission in 1980 will precede the 1984 rendezvous mission, and that
this flyby mission will provide initial data on the physical
nature of the comet and especially the nucleus. More detailed
observation of the nucleus is considered one of the principal
objectives of the rendezvous mission.

As a physical basis of comet phenomena to be observed by
the spacecraft, we adopted the core-mantle model of the evolution of the cometary nucleus, as described by Sekanina[10].
Owing to intense heating of the surface of the nucleus during
possibly thousands of approaches to the sun, an icy envelope,
originally of considerable thickness, gradually sublimates; the
radius of the nucleus shrinks; and after some time the underly-
ing, nonvolatile core becomes exposed to the direct effects of
solar rays. In the subsequent development, molecular desorp-
tion from the unprotected core's surface replaces free sublimation in producing the comet's atmosphere, with transfer of
volatiles from the core's interior to its surface being provided by activated diffusion. The ability of the nucleus to
regenerate sufficient icy materials at the surface is weakened
gradually with time, and finally the whole reservoir of volatiles is exhausted completely. According to this model, the
comet ultimately becomes a "dead" body, i.e., an asteroid.

Long-term declines in the magnitude of Encke and in its
nongravitational forces suggest that the volatiles of the
nucleus available for emissions largely have been depleted.
Relative absence of emitted volatiles after perihelion also
suggests that any symmetric ice crust that once may have existed has been exhausted so that the nucleus is presently a
stable porous object. Accumulated gases that may have migrated
from an icy core to the surface apparently evaporate in sufficient quantity to supply the coma only on the inbound portion
of the orbit. Estimates of the size of a stable nucleus, based
on assumptions of its albedo, give a diameter of 1.3 to 8 km.

Encke's appearance in over 50 observations has depended
on its heliocentric and geocentric distances and has varied

from pass to pass but has displayed most of the major features of comets. A bright center of the coma has been resolved, generally after the "stellar" coma has been in view first. This central condensation is usually recovered by the time Encke reaches 1 a.u. A gas tail (type 1) has been observed, although not on every pass. After perihelion, Encke is faint and diffuse, when visible at all. One important feature is customarily absent or minimal at Encke: a detectable dust component of the coma and tail.

In tailoring the exploration strategy to these characteristics, a mission profile was adopted where the spacecraft arrives at the comet prior to its perihelion passage, at a time when the outgassing process is most active. However, rather than approaching the nucleus immediately and risking possible damage, the spacecraft spends an initial period of 30 to 40 days in a slow traverse of the coma and tail region. During this traverse, navigation data on nucleus location can be obtained conveniently by direct observation from the spacecraft and used in executing accurate final approach maneuvers. The initial approach to the comet therefore can be performed with modest guidance accuracy; hence onboard navigation requirements are simplified greatly.

The Earth-to-comet transfer trajectory is a highly eccentric orbit with an aphelion distance of about 3 a.u., designed to minimize low-thrust propulsion requirements. Typically, the trip time to rendezvous is 800 to 900 days for a mission launched in 1981.

The three-axis controlled electric propulsion spacecraft launched by a Titan IIIE/Centaur booster, has an initial gross mass of 1400 kg and carries 60 to 100 kg of scientific instruments. The required electric propulsion power is 13 kw at Earth departure. A representative spacecraft configuration is shown in Fig. 1. The vehicle consists of a centerbody that houses the equipment and payload compartment, the electric propulsion module, and the mercury propellant tank. A pair of rotatable solar array panels with a wing span of 43 m provides the required propulsive power. This configuration has evolved from a number of previous conceptual design studies[11-13].

Use of electric propulsion is essential to accomplishing this mission, since the characteristically large specific impulse of ion thrusters reduces the amount of propellant required for the comet rendezvous by an order of magnitude compared to chemical propulsion. Thus a mission with adequate payload capacity, launched by a Titan class booster, is made

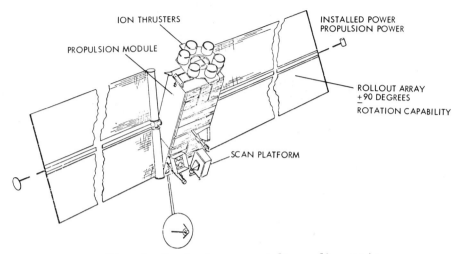

ION THRUSTERS

PROPULSION MODULE

INSTALLED POWER
PROPULSION POWER

ROLLOUT ARRAY
±90 DEGREES
ROTATION CAPABILITY

SCAN PLATFORM

Fig. 1 Typical spacecraft configuration.

feasible. An added advantage is the large maneuver capacity
for excursions through coma and tail following the rendezvous
which is provided by only a few extra kilograms of mercury
propellant.

2. Physical Characteristics of Comet Encke

The Icy Conglomerate Model of the Nucleus. Phenomena observed
in the formation of the coma and tail, and the available evidence of emission of volatile and nonvolatile constituents from
the nucleus, have led to the formulation of the "icy conglomerate model" by Whipple[14] and its more recent extension by
Marsden and Sekanina[15,16]. The nucleus is viewed as an icy
conglomerate of meteoric matter mixed with or containing a
mantle of frozen gases, mostly water and ice or clathrate components. Whipple showed that the nongravitational acceleration
of comets can be accounted for on the basis of mass loss from a
rotating nucleus of this structure.

The core-mantle model, previously discussed, explains the
evolution of the comet nucleus by gradual sublimation of the
icy crust. Figure 2 schematically illustrates the evolution

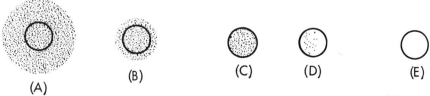

(A)   (B)   (C)   (D)   (E)

Fig. 2 Typical evolution of icy conglomerate nucleus[10].

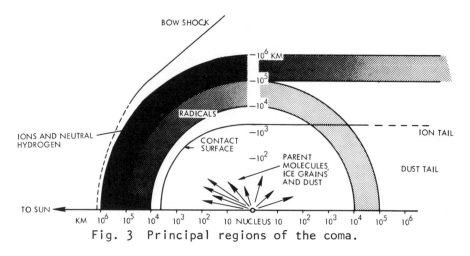

Fig. 3   Principal regions of the coma.

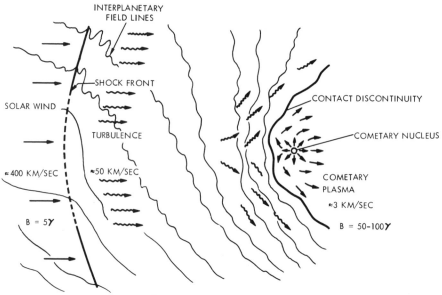

Fig. 4   Solar wind interaction.

process in five stages. Shaded areas show the distribution of
ices. The empty area marks the presence of nonvolatile material.
Encke is presumably in stage C or D at present.

Sekanina[10] developed an analytical model relating the mass
loss to the nongravitational forces. According to this analysis,
the mass loss rate for a typical comet is of the order of 0.01
to 1% of the total mass per revolution. For Encke, the average
mass loss rate during the past 40 years is estimated to be 0.03

to 0.7% per revolution. Marsden and Sekanina[15] give the rate
as 0.03% for the 1967 pass.

Overall Structure of the Coma. As gases, assumed to be princi-
pally water vapor, are emitted from the nucleus, they are photodissociated continuously to form first radicals then ions. A
sufficient number of ions probably have been formed by 10^3 to
10^4 km from the nucleus to make up an outflowing energy density
equal to that of the solar wind's magnetic field, so that a contact surface of accumulated field is formed. Inside this surface, the solar wind and solar wind-accelerated cometary ions
cannot penetrate. Undissociated and un-ionized gases flow outward through the contact surface and ultimately break down to
form radicals at distances from 10^4 to 10^5 km, and then ions,
with no neutrals left at a distance of 10^6 km. The ions are
picked up by the solar wind magnetic field and swept downstream.
The result of this multiple-stage process of coma formation is
shown schematically in Fig. 3, as a series of "layers" in which
various proportions of dissociation products predominate. A
rendezvous mission should penetrate all layers.

Solar Wind Interaction. The supersonic solar wind plasma ultimately must slow down and flow around the comet's contact discontinuity at subsonic speeds (see Fig. 4). One version of the
interaction places a collisionless shock front upstream at a
distance of around 10^6 km, where a drastic reduction in solar
wind velocity takes place. A second, gradual process of solar
wind deceleration also occurs as the relatively heavy cometary
gases become ionized and are picked up by the interplanetary
field and convected downstream, adding their mass to the flow.
Another version of the interaction, not shown, dispenses with
the shock altogether, with the modified solar wind achieving
subsonic flow by ion exchange before reaching the contact surface. In either version, the interaction should produce plasma
instabilities accompanied by electrostatic and electromagnetic
wave turbulence, probably in complex ways not observable around
other bodies in the solar system. The contact surface is itself
subject to instabilities that would generate noise and permit
mixing of cometary and solar wind gas.

Characteristics of Encke Compared with Other Comets. The value
of a selected comet mission depends heavily on whether the
targeted comet is likely to provide information on comets as a
class. Although all comets are to be regarded as individuals,
Encke is reasonably representative of others in dimension and
composition. Figure 5 shows how certain characteristics of
Encke compare with the range of these characteristics shared
with comets in general.

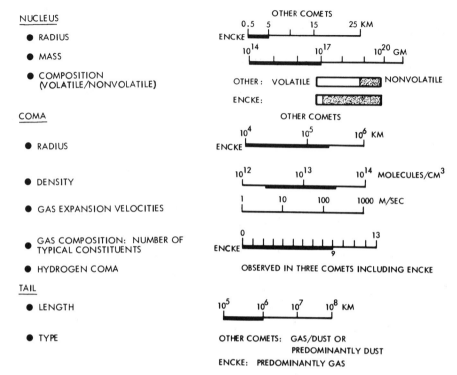

Fig. 5 Encke characteristics compared with those of other comets.

Dimensions of Encke's Features. Like those of most comets, the dimensions of Encke are conjectural at best. The only values that can be attached to individual features are limits or ranges based on apparent sizes obtained with varying observational difficulty. Part of the uncertainty stems, of course, from the intrinsic variability of the coma and tail. Only the nucleus may be thought of as having a definite size at all.

Roemer[17] found that the brightness of Encke's nucleus gave a value of 0.24 for the product of albedo and radius squared, based on an asteroidal brightness law. The resulting dependence of radius R_N on assumed albedo is shown in Fig. 6. Assumption of a geometric albedo of 0.1[15] leads to a radius of 1.8 km. With a density of 1 g/cm³, this leads to a mass estimate of 2×10^{16} g. When the uncertainty of the radius is combined with the uncertainty in density, the uncertainty range of mass becomes large. A low density, typical for a highly porous structure, e.g., 0.1 g/cm³, and a high albedo would lead to a mass of about 10^{14} g. A high density and low albedo would give 5×10^{17} g. For a mass of the order of 10^{16} to 10^{17} g, the acceleration of gravity on the surface is 0.025 to 0.25 cm/sec².

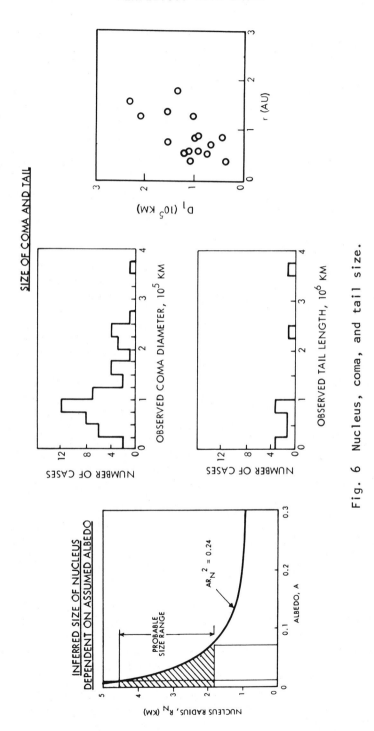

Fig. 6 Nucleus, coma, and tail size.

In the center of Fig. 6 are histograms of all of the values of Encke's coma diameter D_1 and tail length given by Vsekhsvyatskii[18]. The bulk of observations (65%) have given an observable coma diameter of 25,000 to 125,000 km. The most probable range of diameters if 75,000 to 100,000 km. This range is relevant to the rendezvous mission because the coma size depends on solar distance, and it is at the rendezvous distance ($\lesssim$ 1 a.u.) that the coma commonly is measured. All but one of the values above 1.5×10^5 km were observed when the comet was beyond 1 a.u. The well-known dependence of D_1 on solar distance for Encke is shown at right. The histogram of tail lengths (bottom center) shows that the total range of measured lengths is extreme, but most estimates have been below 10^6 km. The highest values were obtained when annual sunspot numbers were over 60.

Chemical Composition.   Indirect information concerning the chemical composition of the nucleus may be obtained from emission spectra of the coma and tail. For Encke, strong lines of CN, C_2, and C_3, with weak lines of CH, NH, OH, CO^+, and N_2^+, are observed. These compounds are almost certainly not constituents of the nucleus but derived by dissociation of compounds such as H_2O, NH_3, CH_4, CO_2, and possibly more complex molecules. Although it is not possible to infer the abundance or even the exact nature of these parent molecules from the spectral data, the spectra indicate that the abundant elements of the C, N, O group played a major role in the condensation and accretionary processes leading to the formation of the cometary nucleus. Hydrogen is present at least insofar as it combined to form compounds such as H_2O, NH_3, and CH_4, and as indicated by hydrogen Lyman-α emission observed surrounding Encke. It is likely that volatile compounds such as CH_4 are trapped as clathrate compounds. Helium and the other inert gases probably were depleted, whereas the lithophilic elements Mg, Fe, Si, Ca, etc., probably were present in something like their solar abundance relative to the CNO group. This assumption leads to the conclusion that about 20% (by weight) of the cometary nucleus consists of oxidized compounds of these elements.

Volatile and Nonvolatile Constituents.   The nucleus of Encke begins to emit sufficient quantities of volatile material to produce a visible coma at a distance of about 1.5 a.u. The total emission of gas per perihelion passage is at present 0.03% of the mass[15]. Using the value of the albedo adopted in this reference, this corresponds to a mass loss of 6×10^{12} g per perihelion passage, or an average loss of $\sim 6 \times 10^5$ g/sec during the approximately 100-day active phase of the perihelion passage, primarily prior to perihelion.

The nucleus core, in its present stage of evolution, consists primarily of nonvolatile material. This material is now in the form of large aggregates, possibly a single piece, and is not swept along readily with the escaping gas. If Encke has a radius of 1.6 km, a mean density of 1.0 g/cm^{-3}, and emits 10^{29} H_2O molecules/sec^{-1} at a velocity of 500 m/sec^{-1}, then the maximum radius of grains with density 1.0 g/cm^{-3} which can be blown away from the comet is $\sim$ 2.5 cm, assuming a drag coefficient of unity. The absence of continuum radiation in the coma of Encke suggests that the coma wind is too weak to blow any significant quantity of solid material away from the nucleus, or the supply of sufficiently small grains in the surface layers has been exhausted.

Although the observed low continuum radiation still permits as much as 10% of the emitted matter to be in the form of grains a few millimeters in radius and an even greater amount in the form of larger particles with a lower ratio of surface to mass, there is no evidence that such particles are now in the coma, and it seems plausible that the fraction of volatile material lost is considerably greater than the fraction of nonvolatile material. A figure of 6 x 10^4 g/sec during the active phase can be used for the rate of emission of meteoric matter, uncertain by of at least an order of magnitude.

The Icy Halo Model. Delsemme and Miller[19] have introduced a model of emission of icy grains which form a bright halo surrounding the nucleus. With this model, the total mass of photodissociated and ionized gases in the coma can be accounted for more readily than with emission of neutral gas from the nucleus only. The icy grains are detached from the snowy surface of the nucleus and accelerated by the emitted gas. While the emitted gas becomes photodissociated and ionized, the icy grains sublimate and release additional gas and trapped radicals. The icy halo thus forms, in effect, a nucleus of enlarged diameter. The presence of this halo may explain the bright central condensation in the inner coma. This model has implications with respect to every cometary feature but is still too new to have been evaluated fully in all of its ramifications.

3. Scientific Objectives and Measurements

Scientific Priorities. Many cometary specialists concur in the view that acquiring information about the nucleus should be the primary goal of a comet rendezvous mission. The nucleus, after all, comprises all of the comet during most of its lifetime and is the source of the more familiar, secondary features the comet displays on approaching the sun. Moreover, it is the nucleus

whose substance and structure may offer clues to the material
and dynamics of solar system formation and to the ultimate formation of an extinct comet. Not all cometary nuclei can be ap-
proached or observed easily, however, because of potentially
hazardous dust, dense snow, or high momentum of expelled gas.
In the case of Encke, the nucleus is more accessible to close
observation, in general, because of the comparably low dust-
emission rate and particularly because of the subsidence in
nucleus emissions as the comet approaches perihelion.

However, emphasis on nucleus observation should not down-
grade the importance in this mission of a balanced set of measurements of Encke's other characteristics. Fortunately, the
versatility of a low-thrust rendezvous mission profile permits
an exploration strategy that is tailored to observation in
depth of all physical features of the coma. In fact, since coma
and tail can be explored initially at smaller risk than the
nucleus in its active preperihelion phase, the exploration strategy discussed in the next section envisions rendezvous with the
nucleus as the final phase of the nominal mission profile.

Classes of Observable Features and Their Priorities. The features of the comet nucleus may be divided into elementary astrometric and physical or chemical characteristics, including detailed composition and structure. These may, in turn, be divided into lithic, nonvolatile or icy, volatile components.

The coma comprises the neutral inner coma, possibly including the icy halo, the ionized coma, and the vast hydrogen cloud,
or extended coma, which reaches out to 10^6km from the nucleus.
A list of these features, grouped into six classes of phenomena
in decreasing order of priority, is given in Table 1. The first
three classes are felt to rank almost equal in priority, all
considerably higher than class 4. The highest priority is as-
signed to characteristics that are least known.

Measurement, Objectives, and Techniques. The following paragraphs discuss considerations involved in selecting a preferred
instrument complement and describe measurement objectives and
techniques. Considerations of mission economy require that some
observation objectives be excluded from implementation, notably
any in situ nucleus observations that would require a lander
package. This means that complex observations, although ranked
high in priority (see Table 1), must be deferred in the interest
of keeping mission plans for the proposed first comet rendezvous
on a realistic basis.

Remote Observation of the Nucleus. Characterization of the
nucleus remains the principal objective of the Encke mission,

Table 1   Classes of observable features in decreasing order
          of priority for 1984 Encke rendezvous

1) External physical characteristics of nucleus	Size Rotation Appearance of details Phase function Temperature	Mass Shape Albedo Fine scale texture

2) Structure and composition of nucleus: nonvolatiles	Surface material
	  abundance
	Subsurface thermal
	  & electrical con-
	  ductivity
	Magnetic proper-
	  ties
	Expelled particle
	  composition
	Internal structure
		Subsurface tempera-
		  ture
		Internal thermal
		  & electrical con-
		  ductivity
		Surface chemical
		  composition
		Expelled particle
		  size, velocity, &
		  spatial distribu-
		  tion

3) Composition of nucleus & inner coma: volatiles	Flux, velocity, composition, & density of neutral gases	
4) Coma formation	Flux & spatial distribution of radicals	Flux & spatial distribution of ions
5) Solar wind interaction & tail formation	Radical & ion spatial distributions Spatial modification of solar wind magnetic field	Spatial dependence of flux, velocity, density, & composition of modified solar wind Occurrence & distribution of plasma & electromagnetic wave modes
6) Extended coma	Size and shape of hydrogen cloud	

subject to the constraints mentioned. Unfortunately, almost
the whole of class 2 properties of the nucleus requires an extremely close approach or even an actual landing on the nucleus.
Observation of these properties therefore simply must be omitted
in defining a realistic set of objectives and instruments. Any-

way, since the nucleus is a completely unknown and uncharted object, a lander package hardly can be justified when even the most elementary properties to be found by the lander are undetermined at the outset.

In contrast, the external astrometric properties of the nucleus are measurable from a distance, and their observation should be a primary mission objective. The items of class 1 are vital features, none of which has been measured directly to any accuracy. The objective is made more feasible by the numerous properties of the nucleus which can be recorded by a relatively few instruments, especially by a TV imaging system. A system capable of resolving features of dimension 0.1 the diameter of the nucleus would be adequate for the scientific imaging requirements.

Although the material of the nucleus will be essentially inaccessible while on the surface, some of it will be expelled and thereby amenable to measurement at a distance. The most promising method is analysis of neutral gases by means of mass spectrometry in the inner coma. Such analysis, since it bears on the question of composition of nucleus constituents, is next in importance to remote measurements of the nucleus itself.

Along with the gases emitted by the nucleus, there should be solid particles, both lithic nonvolatiles and icy grains. If the spacecraft is not moving too quickly, the impacts from the emissions should be of relatively low speed and thus should not cause much physical damage. However, the same characteristic hinders composition measurement of solid particles by conventional means, such as impact ionization mass spectrometry, since the spacecraft's relative motion is much too slow. Compositional data on solid particles can be acquired by detectors carried on a fast flythrough rather than a rendezvous mission. In any case, careful use of the TV, a photometer, an optical particle detector, and a polarimeter (allowing the instruments to look in a direction other than the nucleus) can record the physical characteristics of the solid debris.

The Neutral Coma. The neutral gases emitted by the nucleus are of interest not only as material stemming from the nucleus but also as parent material of the visible coma. It is desirable to measure their density distribution, wind speed and direction, and composition. Some attention should be given to measuring time variations in these quantities, including sudden brightening and the more gradual changes associated with the heliocentric motion of the comet. Regarding atoms and free radicals, it will be necessary to make IR, visible, and UV observations of

the emission and absorption of light. Parent molecules of atoms
and radicals can be examined in more detail by a mass spectrometer without distortion by wall effects. A mass spectrometer
must measure effectively the flux of a given constituent from a
given direction. In order to measure the unknown supersonic
speed of the molecules, and hence convert flux measurements
into density measurements, use should be made of the motion of
the spacecraft, as this will produce measurable aberration of
the neutral molecules.

The Ionized Coma, Contact Surface, and Tail. It is believed
that a relatively dense "ionosphere" surrounds the cometary
nucleus, bounded by a contact surface on the upstream side and
flowing away as a type 1 tail on the downstream side. Appropriate experiments include multichannel photometers, retarding
potential analyzers, or Langmuir probes to measure the electron
density and temperature, ion mass spectrometers to measure the
composition of the plasma, and an orientable Faraday cup or
similar device to measure the flow speed of the plasma. The
electron content of the cometary plasma along the line of sight
to Earth is probably too small for an RF propagation experiment
to give useful data, but an RF plasma resonance experiment may
be feasible as a means of determining density and composition.
As to the neutral and ionized components of the coma, it should
be possible to make use of UV and visible-light spectrophotome-
ters, which have proved successful in Earth's upper atmosphere
and ionosphere.

     Magnetic fields play a vital role in the interaction of
the ionized coma with the solar wind. The most important measurement to be made with a magnetometer, apart from supporting
the solar wind measurements, is to determine the manner in which
the interplanetary magnetic field penetrates the ionosphere of
the comet. It is necessary to determine whether, in fact, a
contact surface exists, and the extent to which it is stable.

Solar Wind Measurements. It is important to determine characteristics of the solar wind flow around the comet, and in particular to determine 1) the strength and configuration of the bow
shock, and 2) the shape and presence of the contact surface.
The characteristics of the flow will be somewhat similar to
those of Earth's magnetosheath (i.e., low Mach number), but the
presence of substantial quantities of singly ionized ions of
cometary origin will make observation more difficult. A Faraday
cup or electrostatic analyzer should be adequate to measure the
gross features of the plasma flow. Determination of the composition of the interaction-influenced solar wind will necessi-
tate the use of more sophisticated instrumentation, perhaps a
crossed-field spectrometer of the type flown on several recent
IMP spacecraft.

Selected Payload Complement.   Table 2 lists the instrument complement proposed for the Encke rendezvous mission and the physical characteristics to be measured by each instrument. Most
of the more complex instruments in this set are seen to be applicable to observation of several characteristics, some to
characteristics of both nucleus and coma. The total weight of
the instruments is estimated as 52 kg, well within the projected
payload weight capacity of the solar-electric rendezvous spacecraft. Most of the instruments listed are those recommended at
the Cometary Science Working Group Meeting in 1971[2].

The list of instruments splits into two groups: those to
be mounted in a fixed position and orientation, and those re-
quiring a scanning capability. The fixed-position instruments
include the optical particle detector, magnetometer, plasma
wave detector, and Langmuir probe. The scanning group is divided into detectors that will predominantly follow the nucleus:
the TV camera, photometer, IR radiometer, mass spectrometer,
and microwave altimeter; and detectors that will scan the coma
or interaction region as well, i.e., the UV radiometer, photopolarimeter, mass spectrometer, and plasma probe. This set of
instruments will be mounted on a two-axis scan platform as shown
in the spacecraft configuration (Fig. 1). This platform can be
time-shared in the sense of serving nucleus-pointing and coma-
pointing needs on a part-time basis.

4. Mission Profile Options and Exploration Strategy

Questions that arise in defining a preferred comet exploration strategy include the following:

1) What type of transfer trajectory and what arrival time is
preferred?

2) How long should the spacecraft remain at the comet after
arrival?

3) What excursion patterns and maneuver sequences are most effective for achieving the comet exploration objectives?

4) What are the preferred operating modes and pointing directions of the scientific instruments?

5) How can approach navigation and guidance requirements be
simplified?

Table 2   Scientific payload complement for 1984 Encke rendezvous

Instrument	Property to which applied
TV image (100 μrad resolution)	Size of nucleus Rotation of nucleus Shape of nucleus Appearance of details of nucleus Size of halo Shape of halo Size of coma Shape of coma Size of tail (uncertain) Shape of tail (uncertain)
Multichannel white light photometer	Albedo of nucleus Phase function of nucleus Albedo of halo Phase function of halo Brightness profile of halo
IR radiometer	Temperature of nucleus
Photopolarimeter	Fine scale texture of nucleus Fine scale size distribution of ice grains of halo Fine scale size distribution of non-volatile particles of coma
Microwave altimeter	Mass of nucleus Size of nucleus Surface composition of nucleus
Radiometer, UV 1000-4500Å	Distribution of ionized gases in coma, contact surface, and tail
Optical particle detector (Sisyphus)	Distribution, velocity of icy grains of coma Distribution, velocity of nonvolatile particles of coma
Mass spectrometer	Flux, velocity, density, spatial distribution of neutral and ionized gases of coma
Magnetometer	Magnetization of nucleus Magnetic field configurations of contact surface, tail, and interaction region

Table 2 (Cont'd)

Plasma wave detector   Electric waves in contact surface, tail,
                              and interaction region
                            Local electron densities in· ionized coma

Langmuir probe   Local electron densities in ionized coma

Plasma probe   Flux, density, energy spectrum of solar
                              wind and reduced solar wind in inter-
                              action region

Estimated mass of optical detectors, kg	33.0
Estimated mass of altimeter, kg	6.0
Estimated mass of gas & plasma property analyzers, kg	13.0
Total mass, kg	52.0

6)   How can the spacecraft be protected against the adverse
thermal environment and against hazards due to the flux of
cometary particles?

7)   How much flexibility is needed to adapt the mission profile
to unforeseeable conditions?

     Basic characteristics of the three main mission phases,
1) Earth-to-comet transfer, 2) comet approach, and 3) comet exploration following the rendezvous, will be reviewed briefly in
the following paragraphs. Criteria for selection of preferred
operating modes include: effectiveness in achieving the scientific mission objectives, simplicity of system implementation
and operation, cost economy, maximum use of conventional processes and available technology, and limited exposure to environmental hazard.

Transfer Trajectory Characteristics. Typical low-thrust transfer trajectories for a 1984 Encke rendezvous are illustrated in
Fig. 7. A common arrival date, 50 days before the comet's perihelion passage, is assumed in these examples. Large aphelion
distances (R_A) and corresponding long flight times are typical
for these transfer trajectories. Since the spacecraft velocity
must be matched to the comet's velocity at arrival through continuous thrusting during transfer, the trajectory profiles shown
are advantageous in minimizing the required thrust level (i.e.,
electric propulsion power) as well as propellant mass. Thrust
orientations required at different phases of the transfer orbit
are indicated in the graph.

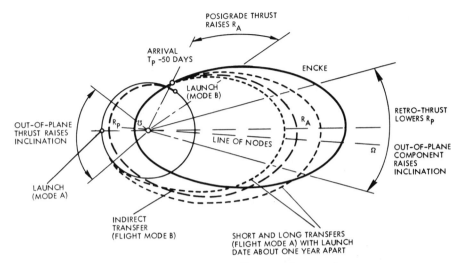

Fig. 7 Typical transfer trajectories.

Two trajectory types are identified in Fig. 7: direct trajectories (flight mode A), which depart Earth in an <u>outbound</u>
direction, with launch dates occurring near the longitude of
the comet's perihelion, and indirect trajectories (flight mode
B), which have a more flexible launch date and generally depart
Earth in <u>inbound</u> direction. Direct trajectories can be launched
at dates about 1 yr apart. Typical trip times are 700 and 1050
days. Indirect trajectories can be launched at a wider range
of longitudes at dates that complement the launch dates for
direct flights. Trip times range from 800 to 1000 days.

An overview of possible transfer trajectory options for a
spacecraft, launched by the Titan IIIE/Centaur and using 15 kw
of solar-electric propulsion power at 1 a.u., is provided by
the mission maps shown in Fig. 8. The map shows contours of
net spacecraft mass (the mass remaining after the solar array,
electric propulsion hardware, and propellant mass are subtracted
from the initial gross mass) in a plot of launch date vs arrival
date. Diagonal lines indicate flight time. All data points re-
flect payload performance achieved by an optimal electric thrust
program. Payload performance for power levels other than 15 kw
can be determined by proportional scaling, i.e., changing the
indicated net spacecraft mass in the same ratio as the reference
power level.

The following conclusions regarding possible transfer trajectory options are derived from the characteristics shown in
the mission map:

1) Two classes of trajectories are available: the slow trajectories (on the left) deliver significantly larger maximal

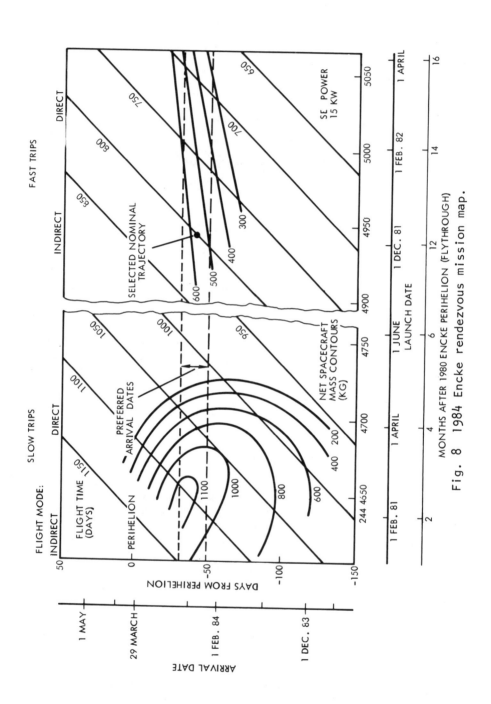

Fig. 8 1984 Encke rendezvous mission map.

payloads than the fast trajectories (on the right). Since a
net spacecraft mass of 400 to 500 kg is adequate for the mission, a fast trajectory with a trip time of 750 to 800 days can
be selected.

2) A short flight time is preferred, not only because it reduces thrust time and hence the probability of propulsion failure, but also because it permits a much later launch date.
Thus, the results of a precursor mission to comet Encke contemplated for 1980 could be utilized more effectively in preparing
for the 1984 rendezvous. A second time scale at the bottom of
the mission map shows the time elapsed from the 1980 perihelion
passage of Encke. The fast trip options allow 8 to 10 months
more lead time than the slow trip options.

3) The preferred arrival time is 30 to 50 days before perihelion passage, in accordance with scientific mission objectives.
A horizontal strip shown in the mission map brackets these arrival dates. We note that a 25-to-50-day launch date variation
can be accommodated readily in the fast trip option without af-
fecting the payload mass if the arrival date is changed by only
a few days.

The choice of arrival time is dictated in part by the comet
exploration strategy. Rendezvous 40 days before perihelion will
provide adequate time for extended coma and tail exploration,
when the comet is most active and permits arrival at the nucleus
about 10 to 20 days before perihelion, following coma and tail
exploration, as will be discussed below.

On the basis of these considerations, a nominal transfer
trajectory with these characteristics was selected: launch
date, Dec. 8, 1981; arrival date, Feb. 16, 1984 (40 days before
perihelion); distance from sun at arrival, 0.95 a.u.; flight
time, 800 days; total thrust time, 751 days; net spacecraft
mass, 527 kg (at 15 kw propulsion power); and departure hyper-
bolic excess velocity, 8 km/sec.

Comet Approach Phase. The approach trajectory relative to the
comet for the final 100 days of transfer is shown in Fig. 9.
During this time period, the thrust vector is oriented almost
directly opposite to the line of sight from spacecraft to comet.
This means that an optical navigation sensor carried by the
spacecraft must look essentially along the thrust beam in order
to observe the comet and with possible field-of-view obstruction
by the spacecraft body. Thus, intermittent reorientation of
the spacecraft will be necessary to permit an unobstructed view
by the navigation sensor, probably accompanied by thrust inter-
ruption.

Relative Motion in Comet Vicinity. Exploration of the coma and
nearby tail regions is performed on a trajectory that includes

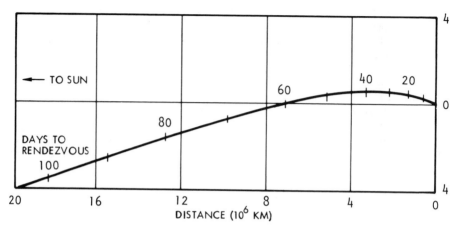

Fig. 9 Final approach to rendezvous.

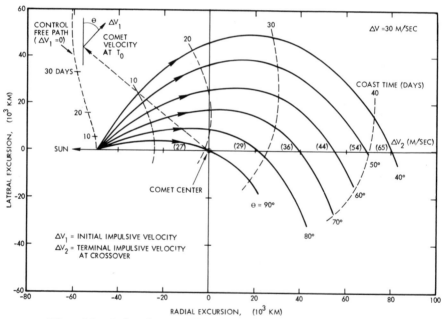

Fig. 10 Relative trajectories with initial offset ΔR_o = -50 x 10 km; $T_o = T_p$ -50 days.

alternating coast and thrust phases. Figure 10 shows a set of coast arcs in cometocentric coordinates, originating from a 50,000-km offset. The start time is 50 days before perihelion passage, and the initial velocity increment (ΔV_1) is 30 m/sec. The trajectories are in the plane of the comet's motion around the sun. The comet's heliocentric velocity is indicated by the slanted vector pointing to the upper left. Elapsed coast time

is indicated by parametric lines. Local velocities (ΔV_2) at the
abscissa crossing points are given by numbers in parentheses.

   The curved character of the relative trajectories is due
to 1) the Coriolis effect in the rotating coordinate system
adopted here, and 2) solar differential gravity. A dashed curve
at the left shows the drift due to solar differential gravity
in the absence of an initial velocity increment. Nucleus gravity is negligible at the distances considered here. Corresponding coast arcs running in opposite direction would be described by a similar (antisymmetric) set of trajectories.

   Coast arcs of this type can be used to synthesize a coma/
tail exploration pattern, starting at an initial offset point
on the sunward side and arriving at the comet center after an
elapsed time of 30 days. Typically, the total velocity increment for this excursion is 200 to 300 m/sec. Reduction of the
excursion distance or increase in excursion time would reduce
the velocity requirement.

Perturbing Forces and Stationkeeping Requirements. The principal perturbing influences acting on the spacecraft while stationkeeping near the nucleus are: 1) solar differential gravity,
2) solar pressure, 3) gas flow pressure, and 4) nucleus gravity.
Solar differential gravity is only an apparent perturbation
effect introduced by adopting the cometocentric coordinate system as frame of reference. It is by far the dominant effect
that must be compensated if stationkeeping at a fixed relative
position, more than a few hundred kilometers from the nucleus,
is desired. It varies linearly with distance from the comet
center and inversely with the third power of solar distance.
Because of this dependence on solar distance, the differential
gravity effect is about 25 times larger at perihelion than at
1 a.u.

   Figure 11 shows the different perturbing forces acting on
a 1000-kg spacecraft having a 100-m^2 solar array, as function
of distance from the nucleus. The solar distance is 0.34 a.u.
(perihelion). Forces due to gravity and gas flow pressure, both
decreasing with the inverse square of the distance, tend to can-
cel if the solar array is deployed fully. The maximum gas flow
pressure was derived for an assumed mass flow rate from the
nucleus, 6 x 10^5 g/sec (see Sec. 2). The maximum pressure at
the surface of the nucleus is about 60 mlb but only 1 mlb at a
distance of 10 km.

   Gravity acceleration at the surface of the nucleus
(radius = 1.8 km) is 0.41 x 10^{-3} m/sec^2. Thus the gravity
force that would be acting on a spacecraft hovering near the

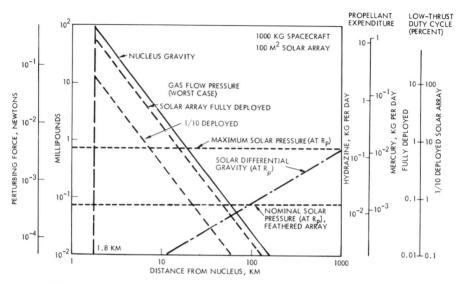

Fig. 11 Perturbation forces and thrust requirements in center of comet.

surface is 90 mlb. The nucleus gravity and solar differential gravity effects are equal at a distance of 30 to 40 km, as shown in the graph. Solar pressure at perihelion is shown for a fully deployed solar array and an array turned away from the sun for thermal protection. It ranges from 0.1 to about 1 mlb. These results show that the combined perturbation effects are smallest in the range of 50 to 100 km from the nucleus. This fact is of potential interest if an extended stationkeeping period in the penumbra of the nucleus for purposes of thermal protection should be desired.

The concept of thermal shielding by the nucleus is actually quite feasible (see Ref. 8, p. 5-27). It can be used, for example, to reduce the thermal load at perihelion by a factor of 2 if the spacecraft remains in partial eclipse for 26 days. Although the position behind the nucleus can be used for scientific measurement, e.g., solar absorption spectra, it constrains freedom of exploration during the most important part of the mission. This mode should, therefore, be considered only in an emergency.

Protection Against Adverse Environment. Regarding thermal protection, the spacecraft must be capable of withstanding the close solar distance during the comet exploration phase. Survival at 0.34 a.u. when the thermal load is 8.65 times greater than at 1 a.u. is a prerequisite to observation of the comet

beyond perihelion passage. The scientific importance of extended exploration beyond this point is sufficiently great to
warrant the required additional design complexity.

Protection of the solar array against intensive solar heating is achieved by deflection from full exposure, starting at
0.38 a.u. Thus the maximum solar cell temperature can be held
below 140°C. After rendezvous at 1 a.u., the spacecraft will
require only intermittent thrust power, at a reduced level, and
some degradation of solar array performance is acceptable. During all thrust and observation phases, the spacecraft body can
be maintained at attitudes that exclude exposure of the thermal-
ly sensitive rear surface.

Protection against particle flux from the nucleus is reduced in importance by arrival at the nucleus at a time (i.e.,
close to the perihelion) when the flux will have subsided
partly. Estimated impact rates of milligram-size particles
range from 10^2 to 10^4 per day near the nucleus surface (see
Ref. 8, p. 6-5). The principal hazard anticipated under the
very low emission and impact velocity (3 to 10 m/sec) of these
particles is due to deposition of low-density particles of
fluffy structure on thermal insulation blankets and exposed op-

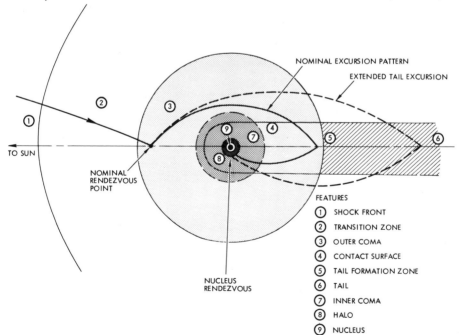

Fig. 12 Excursions through principal comet features
(not to scale).

tical sensor apertures. The best protection is to avoid pro-
longed exposure at distances of less than about 20 to 50 km
from the nucleus.

Selected Nominal Comet Exploration Path. The selected comet
exploration profile is illustrated in Fig. 12. All areas of
primary interest, numbered 1-9, are visited by the spacecraft
in the course of its passage through the coma, nearby tail
region, and to the nucleus. Extension of the path deeper into
the tail (point 6) can be included if Earth observation indicates this to be warranted.

Advantages of this mission profile are summarized as follows: 1) avoidance of exposure to active nucleus on arrival at
comet 40 days before perihelion; 2) thermal protection behind
nucleus is available if necessary; 3) initial guidance accuracy
is not critical for arrival at offset rendezvous point 50,000
km from nucleus; 4) detection of nucleus and navigational fixes
are simplified if postponed past arrival at offset rendezvous
point; 5) residual arrival velocity ($\sim$ 30 m/sec) can be used to
initiate coma/tail traverse; and 6) arrival time at nucleus can
be controlled (and postponed if necessary until emissions sub-
side) on the basis of local observations during coma traverse.

5. Conclusion

The exploration strategy discussed in this paper is for a
specific cometary target (Encke in 1984) allowing at least 80
days of coma, tail, and nucleus observations after achieving
rendezvous prior to the perihelion passage. Such a strategy
can be adapted readily to other comets not as well known as
Encke. The principal trajectory control difficulty inherent
in the large ephemeris uncertainty of most cometary targets is
resolved by the two-stage rendezvous approach, where initial
arrival point dispersions of 50,000 km and residual velocities
of 30 m/sec have no adverse effect on the exploration profile.
Essential to the entire rendezvous mission concept is the use
of solar-electric propulsion, not only during the transfer and
approach phases, but also through the extended exploration phase
where a significant maneuvering capability is desirable.

References

[1] "Comets and Asteroids: A Strategy for Exploration," Report of
the Comet and Asteroid Mission Study Panel, NASA TMX-64677.

[2] "Proceedings of the Cometary Science Working Group," June 1971, Yerkes Observatory, Williams Bay, Wis.

[3] "Proceedings of the NASA Comet Conference," April 8-9, 1970, University of Arizona, Tucson, Ariz.

[4] Brereton, R. G., Newburn, R. L., Giffin, C. E., Neugebauer, M., Smith, E. J., and Willingham, D. E., "Mission to a Comet: Preliminary Scientific Objectives and Experiments for Use in Advanced Mission Studies," Jet Propulsion Laboratory Technical Memo 33-297. Feb. 15, 1967, Jet Propulsion Laboratory, Pasadena, Calif.

[5] Friedlander, A. L., Niehoff, J. C., and Waters, J. I., "Trajectory and Propulsion Characteristics of Comet Rendezvous Opportunities," Rept. T-25, Aug. 1970, Astro Sciences, IIT Research Institute, Chicago, Ill.

[6] Whipple, F. J., "Why Missions to the Comets," Astronautics & Aeronautics, Vol. 10, Oct. 1972, pp. 12-17.

[7] Farquhar, R. W. and Ness, N. F., "Two Early Missions to the Comets," Astronautics & Aeronautics, Vol. 10, Oct. 1972, pp. 32-37.

[8] "Study of a Comet Rendezvous Mission," Final Rept. 20513-6006-RO-00, May 12, 1972, prepared for Jet Propulsion Laboratory, Pasadena, Calif. by TRW Systems, Redondo Beach, Calif., Contract 953247.

[9] Atkins, K. L. and Moore, J. W., "Cometary Exploration: A Case for Encke," AIAA Paper 73-596, July 10-12, 1973, Denver, Col.

[10] Sekanina, Z., "Dynamical and Evolutionary Aspects of Gradual Deactivation and Disintegration of Short-Period Comets," The Astronomical Journal, Vol. 74, Dec. 1969, pp. 1223-1234.

[11] Goddard, F. E., Parks, R. J., Briglio, A., Jr., and Porter, J. C., Jr., "Solar Electric Multimission Spacecraft (SEMMS) Phase A Final Report, Technical Summary," Rept. 617-2, Sept. 10, 1971, Jet Propulsion Laboratory, Pasadena, Calif.

[12] "Study of a Common Solar-Electric-Propulsion Upper Stage for High-Energy Unmanned Missions," Vol. II, July 14, 1971, prepared

for NASA/OART, Advanced Concepts and Missions Division, Moffett Field, Calif. under Contract NAS2-6040, TRW 16552-6007-RO-00.

[13]Horio, S. P., "Solar Electric Propulsion Asteroid Belt Mission Study," SD 70-21-1, Jan. 1970, North American Rockwell Corp., Downey, California.

[14]Whipple, F. L., "A Comet Model I, The Acceleration of Comet Encke," Astrophysical Journal, Vol. 3, 1950, pp. 375-394.

[15]Marsden, B. G. and Sekanina, Z., "Comets and Nongravitational Forces, IV," The Astronomical Journal, Vol. 76, 1971, p. 1135.

[16]Sekanina, Z., "Total Gas Concentration in Atmosphere of the Short-Period Comets and Impulsive Forces upon their Nuclei," The Astronomical Journal, Vol. 74, Sept. 1969, pp. 944-950.

[17]Roemer, E., "The Dimensions of Cometary Nuclei," Mémoires Société Royale des Sciences de Liége, Vol. 12, 1966, p. 23.

[18]Vsekhsvyatskii, S. K., "Physical Characteristics of Comets," TT F-80, 1964, Office of Technical Services OTS 62-11031.

[19]Delsemme, A. H. and Miller, D. C., "Physico-Chemical Phenomena in Comets-II, Gas Absorption in the Snows of the Nucleus," Planetary and Space Science, Vol. 18, 1970, p. 717.

Index to
Contributors to Volume 50